MECHANICS OF FRETTING FATIGUE

SOLID MECHANICS AND ITS APPLICATIONS
Volume 30

Series Editor: G.M.L. GLADWELL
Solid Mechanics Division, Faculty of Engineering
University of Waterloo
Waterloo, Ontario, Canada N2L 3G1

Aims and Scope of the Series

The fundamental questions arising in mechanics are: *Why?*, *How?*, and *How much?* The aim of this series is to provide lucid accounts written by authoritative researchers giving vision and insight in answering these questions on the subject of mechanics as it relates to solids.

The scope of the series covers the entire spectrum of solid mechanics. Thus it includes the foundation of mechanics; variational formulations; computational mechanics; statics, kinematics and dynamics of rigid and elastic bodies; vibrations of solids and structures; dynamical systems and chaos; the theories of elasticity, plasticity and viscoelasticity; composite materials; rods, beams, shells and membranes; structural control and stability; soils, rocks and geomechanics; fracture; tribology; experimental mechanics; biomechanics and machine design.

The median level of presentation is the first year graduate student. Some texts are monographs defining the current state of the field; others are accessible to final year undergraduates; but essentially the emphasis is on readability and clarity.

For a list of related mechanics titles, see final pages.

Mechanics of Fretting Fatique

by

D. A. HILLS

and

D. NOWELL
Department of Engineering Science,
University of Oxford,
Oxford, U.K.

KLUWER ACADEMIC PUBLISHERS
DORDRECHT / BOSTON / LONDON

A C.I.P. Catalogue record for this book is available from the Library of Congress.

ISBN 0-7923-2866-3

Published by Kluwer Academic Publishers,
P.O. Box 17, 3300 AA Dordrecht, The Netherlands.

Kluwer Academic Publishers incorporates
the publishing programmes of
D. Reidel, Martinus Nijhoff, Dr W. Junk and MTP Press.

Sold and distributed in the U.S.A. and Canada
by Kluwer Academic Publishers,
101 Philip Drive, Norwell, MA 02061, U.S.A.

In all other countries, sold and distributed
by Kluwer Academic Publishers Group,
P.O. Box 322, 3300 AH Dordrecht, The Netherlands.

Printed on acid-free paper

Printed in the Netherlands

Contents

Preface

Failures of many mechanical components in service result from fatigue. The cracks which grow may either originate from some pre-existing macroscopic defect, or, if the component is of high integrity but highly stressed, a region of localized stress concentration. In turn, such concentrators may be caused by some minute defect, such as a tiny inclusion, or inadvertent machining damage. Another source of surface damage which may exist between notionally 'bonded' components is associated with minute relative motion along the interface, brought about usually be cyclic tangential loading. Such fretting damage is quite insidious, and may lead to many kinds of problems such as wear, but it is its influence on the promotion of embryo cracks with which we are concerned here. When the presence of fretting is associated with decreased fatigue performance the effect is known as fretting fatigue.

Fretting fatigue is a subject drawing equally on materials science and applied mechanics, but it is the intention in this book to concentrate attention entirely on the latter aspects, in a search for the quantification of the influence of fretting on both crack nucleation and propagation. There have been very few previous texts in this area, and the present volume seeks to cover five principal areas;

(a) The modelling of contact problems including partial slip under tangential loading, which produces the surface damage.

(b) The modelling of short cracks by rigorous methods which deal effectively with steep stress gradients, kinking and closure.

(c) The experimental simulation of fretting fatigue.

(d) The correlation of data from experiments in (c), together with the production of quantified growth criteria.

(e) The development of understanding of surface phenomena known critically to influence fretting fatigue, such as surface finish and friction.

This work has built on studies carried out at Oxford under the supervision of Dr. J.J. O'Connor during the late 1960's, and picked up by the current authors from about 1986. We remain indebted to John O'Connor for his inspiration and continuing stimulation. Our work has been supported largely by the SERC, and more recently by Rolls Royce plc, and we gratefully acknowledge their assistance. A subject of this kind does not stand still, and so whilst we have here tried to lay the foundation stones for those aspects of the subject which we believe to be largely uncontroversial, and which we hope may stand the test of time, we have also included new material which is still being developed. We are at present delving ever more deeply into a quantitative explanation of crack initiation and we expect that this will be the most fruitful area of study over the next decade.

We would like to express our thanks to Dr. Danong Dai for his recent contributions to the eigenstrain method of crack analysis, and to Richard Munisamy for his efforts on the partial slip problem. Professor Graham Gladwell, whose text on contact mechanics (Gladwell, 1980) has stimulated much of our own work, gave generously of his time in helping us to edit this book, and made many useful suggestions. We are indebted to him for his contribution, and for the improvements he has helped us to make. Lastly we would like to thank Anne and Ray Hills for their help with the preparation of the manuscript.

D.A. Hills
D. Nowell
Oxford, January 1994

Chapter 1

Introduction

In the design of any mechanical structural component, two criteria play very important parts. These are: the avoidance of yield, and the avoidance of brittle fracture. It is relatively easy to design against yield; this simply requires the determination of a composite stress parameter from the individual stress components, and this parameter is maintained below the value of the yield stress at all points. Design against brittle fracture is more difficult; this requires not only the determination of the state of stress within an object, but also some speculation on the form, location, and origin of initial defects, which often grow by the process of fatigue during the service life of the component, and may reach a critical value leading to brittle fracture. This latter criterion forms the basis of the study of fracture mechanics. There is now an excellent range of introductory textbooks on this subject, such as those by Gdoutos (1990, 1993), Ewalds and Wanhill (1984), and Broek (1982, 1989). It is important to recognise at the outset that the fatigue life of a crack has two quite distinct elements, and that the relative proportion of the total life expended in each portion will vary tremendously. These two phases are the initiation or nucleation phase, and the propagation stage. In welded structures, or those components made from castings of indifferent quality there will pre-exist a number of defects of appreciable size, due to the presence of inclusions, slag, or lack of weld penetration. In these components there is no initiation time, so that the entire life of the component is expended in propagation. However, most critical components in high-technology applications are not manufactured by these routes, and the surface finish is usually very good. There are therefore no pre-existing macroscopic defects, and it is very difficult to speculate where they may form. Indeed, it is the designer's job to ensure that the geometry of the components is such as to reduce the chances of initiation to a minimum! This is normally done by avoiding stress raisers such as fillets, keyways, screw threads and abrupt changes of section. All these measures are concerned with keeping the stress state as uniform as possible, and are successful when the possible sites of initiation lie on a free surface. However, where there is a connection between two components, such as a collar heat shrunk onto a shaft, another form of crack initiation, which is

Figure 1.1: Close-up photograph of fretting damage on a mild steel shaft used to support steel blocks in a fatigue apparatus.

much more aggressive than that prevalent at a free surface becomes possible. This is the phenomenon of fretting.

Fretting occurs whenever a junction between components is subjected to some oscillating force, and this gives rise to some minute relative tangential displacement over at least part of the interface. Most frequently this condition occurs in bolted or mechanically fastened joints. A typical scar resulting from fretting damage is shown in fig. 1.1. The shaft shown supported a pair of blocks which were notionally static with respect to the shaft itself, and the assembly was part of an experiment mounted in a hydraulic test machine. The fluctuating loads present elsewhere in the test caused very small axial relative displacements under radial load, and, after several millions of cycles of load, produced the damage depicted. In this instance, because the shaft carried its load principally in shear, there was no bulk axial tension present to develop and propel a crack. The damage was therefore relatively benign: in such cases it is not uncommon for the wear products produced by fretting (the 'cocoa' seen on the surface) to be bulkier than the metal from which it was formed. This produces an interference fit between the mating components, and may lead to seizure. In cases where an oscillating bulk tension is present in addition to the contact forces, crack propagation will often result. This phenomenon of 'fretting fatigue' is a frequent source of failure of components which are initially of high integrity. In this book we shall principally be interested in the propensity of fretting to exacerbate fatigue, and little attention is given to fretting wear. For a discussion of that subject, and in particular the influence of surrounding aggressive environments on fretting the text by Waterhouse (1972) should be consulted.

Returning to fretting fatigue, we note that the principal previous text on the subject was also compiled by Waterhouse (1981) where both mechanics and materials aspects of the subject are addressed. That book was subsequently brought up to date in a review paper, (Waterhouse, 1992). Also, two volumes of conference proceedings have been published recently; Attia and Waterhouse (1992), and Waterhouse and Lindley (1993). Bolted flange connections between pipes provide a typical example of a fretting fatigue problem. The pre-load in the bolts is sufficient to maintain the flanges in intimate contact and resist the internal hydrostatic pressure. But frequently in such assemblies there are fluctuating loads, perhaps arising from the internal fluid being driven by a reciprocating pump, perhaps from more general sources of vibration. These fluctuations, whilst not themselves causing distress to the material, are sufficient to induce relatively high transitory inertia forces, and the clamping force provided by the bolt ring may be insufficient to inhibit all tangential displacement. The bolts themselves provide only the clamping force, and displacement is resisted only by friction. Thus, the development of a large coefficient of friction may help to generate large tangential forces resisting movement, whilst a well-lubricated joint may suffer large displacement, but with only a relatively small tangential force developed. It is important to recognise that in any joint the area of contact may be composed of a complex pattern of slip regions, where relative tangential displacement of corresponding particles on the two contacting bodies occurs, limited only by the coefficient of interfacial friction, and stick regions. It is the finite elasticity or compliance of the bodies which permits them to slip past each other in one region whilst sticking in another, and we distinguish between the term 'slip' which occurs at a part of an otherwise adhered interface, from 'sliding', which is the bulk displacement of one body relative to another. In problems involving complex load histories, stick at a particular point during one part of the cycle may subsequently give way to slip, and *vice versa.* Indeed part of the purpose of this text is to indicate methods for determining the condition of stick and slip regions. Stick is, of course, equivalent to adhesion, and no surface damage is suffered; in the slip regions the relative movement causes the expenditure of frictional energy, which provides structural damping, but at the same time causes surface degradation and the generation of a small amount of heat. The form of the surface degradation varies considerably according to the material combination, the state of the stress, the amplitude of the motion, the presence or absence of surface treatment and contaminants, the nature of the surrounding fluid/gas, the temperature and possibly even the geometry of the contact itself.

The study of plain fatigue has its origins in the analysis of railway axle failures carried out by Wöhler in the 1880's, but *fretting* fatigue was not studied systematically until 1927 when Tomlinson designed a fretting fatigue machine consisting of contacting annuli in small amplitude rotational oscillation. Even this work concentrated on the damage and corrosion associated with fretting, and it was Warlow-Davies (1941) who examined the influence of fretting damage on fatigue life. In his tests, Warlow-Davies imposed fretting damage first and subsequently conducted fatigue tests; it was left to McDowell (1953) to carry out simultaneous fretting fatigue tests, and subsequent pioneering work was conducted by Fenner and Field (1958) and by Waterhouse. Waterhouse's extensive work was summarised in his books of 1972 and 1981. In the development of our understanding of the allied mechanics, the first major achievement, which is still commonly used, was the analysis of the general Hertzian contact under conditions of partial slip, by Mindlin (1949). Mindlin and colleagues went on to refine and extend the analysis, and their results were used extensively by Johnson (1955) in a careful experimental investigation of the fretting phenomenon. Since then the major advances pre-dating those given in this text were made by Wright (1970) and Bramhall (1973), reported in detail by O'Connor (see Waterhouse, 1981); each involves a greater refinement in modelling of the contacts, and carefully controlled experiments. Bramhall noticed an apparent size effect in fretting, and also pioneered the used of Hertzian contacts (rather than flat-ended complete contacts) to induce fretting damage.

In this book, we will examine exclusively the mechanics of the fretting fatigue process, without specifically attempting to identify material combinations to suppress the phenomenon. The aim of this work must ultimately be to provide design criteria for any pair of contacting bodies, to enable the engineer explicitly to avoid the conditions where a crack may initiate. However, if conditions are so arduous that the initiation of cracks is unavoidable, the influence of the fretting contact on accelerating fatigue crack growth must also be quantified. When once the crack is sufficiently long for the initiation phase of its life to be complete, crack growth is controlled entirely by the crack tip stress intensity factors experienced. These are completely defined by the state of stress, which is composed of the bulk stress state, together the contact stress field. If the crack tip stress intensity factor may be found reliably, the crack growth may be predicted from the Paris fatigue law, and the problem is very little different from plain fatigue, save that the crack exists in a steep stress gradient. Interest therefore centres on the influence of fretting on the initiation and very early development of the crack. What are the variables which control initiation, and what are the mechanics of the process? These questions are not easy to answer, at least partly because questions of crack initiation rely on phenomena which occur within a single grain, and are not necessarily amenable to quantification by a bulk analysis. Nevertheless, parameters quantifying the initiation process will be postulated. We will assume that, for a given pair of materials, a knowledge of the stress, strain and displacement fields, including implicitly the surface values, i.e. tractions and relative slip, is sufficient to quantify the process of crack initiation, early development and growth. The physical geometry of the contact will be assumed *not* to affect crack development[1], other than in the way it controls the field variables cited. In a real problem where fretting fatigue is anticipated, such as the flange joint cited earlier, the state of stress will almost certainly have to be determined by a numerical method (normally the finite element method). Also, it may be very difficult indeed to obtain concise expressions for the size and shape of surface slip regions, and indeed to

[1]Note that this is quite different from wear and, in particular, fretting wear. In these cases debris may either be trapped as third bodies or released (often manifested as cocoa) depending on both the shape and orientation of the contact.

describe the traction distribution itself with any rigour. The last difficulty arises in connections of the type described because the contacting points - flanges, gaskets, bolts and washers - are often in the form of complete contacts, i.e. the contact area is determined solely by geometry, and is independent of load. For contacts of this kind, the presence of burrs or machining imperfections can often have a profound influence on the pressure distribution. Also, singularities are often present at the edges of the contact. These are difficult to model, and are relieved by local plasticity. For these reasons, experimental work done by the authors has concentrated on the use of a well-defined geometry, of the Hertz type. Experiments of this kind are much more difficult to devise than those using a 'bridge' specimen with flat-ended feet, and a critical review of the types of tests available will be given in Chapter 7.

There are certain characteristics which appear to be common to the cracks generated by a wide range of fretting geometries. First, cracks invariably nucleate at the edge of the contact, as this is where the tension induced by the oscillating contact is highest (if there is a gradient in the bulk tension field, cracks start on the side sustaining the higher tension). Secondly, the early growth is in mode II crack loading (see §6.1), so that the crack extends in a direction of maximum shear. Adjacent to the free surface these directions lie at $\pm$ 45 degrees to the surface, and so the embryo crack has two possible trajectories. In practice cracks invariably grow *beneath* the contact. At some critical length the crack front turns to an orientation giving the maximum mode I stress intensity, which, as it normally occurs at a depth comparable with the semi-width of contact, means that the state of stress is dominated by the bulk loading. The new trajectory is therefore normal to the surface, and the crack propagates at an ever faster rate until fracture occurs. As it is very difficult to measure crack growth rate beneath a fretting contact it is usual to assess fretting fatigue performance by measuring the total life of the specimen.

The life of a specimen (and therefore the life of a component in a prototypical application) depends on a number of characteristics, which the experiments are designed to elucidate, but on which we may briefly speculate. To initiate a crack, we require a reversing state of shearing stress, in order to provide the environment for dislocation migration, to form a slip band. The more severe the reversing state of stress the more likely a slip band is to form, and therefore reversing contact conditions which give rise to severe surface stress states are the ones which are most likely to initiate a crack; this means that a high coefficient of friction, together with a large tangential force (which is, of course, ultimately limited by friction), together with a concentration of contact pressure (for example induced by a square-ended contact pad) produce severe conditions. The presence of a bulk tension parallel with the surface is also known to encourage crack development; this is presumably because it helps to keep the crack faces open, so assisting in the influence of shear in inducing mode II loading, and also because it will stabilise and propagate a mode I crack further on in the crack's life. It is often remarked that the amplitude of relative tangential displacement of the contacting pad has a profound influence on fretting performance: at low amplitudes the fretting influence is small. It increases with relative displacement up to some particular value, and then decreases at large amplitudes. It is usually accepted that the lack of effect at large displacements is because fretting wear is taking place, so that any embryo crack is ground away. At lower values of relative displacement, the reduction in fretting fatigue life with increasing amplitude may originate from two sources: first, because for partially slipping contacts the relative tangential displacement is directly related to the magnitude of the shearing force, and secondly because in sliding mode the amplitude of displacement controls the number of load reversals on an asperity scale (see §9.4).

In the previous paragraph we have introduced briefly the *mechanical* attributes of the system which might be expected to control the fretting fatigue performance. The question arises as to how material properties might affect the problem. First, initiation is concerned with plasticity on a very small scale, within an individual grain, so that the stress needed to cause a dislocation to glide within a grain is a key quantity, best characterised by the friction stress. This is not a true material property, as it depends on the Peierls-Nabarro force (the force needed to drive a dislocation through an otherwise perfect, infinite lattice), together with the influence of grain size, morphology, and dislocation interaction. When once the crack has initiated, the quantities controlling both crack growth and catastrophic failure are precisely the same as those relevant to plain fatigue, i.e. the relationship between stress intensity factor range and crack growth rate, and the fracture toughness (see §6.1). Thus, in general the material properties which hinder crack growth in plain fatigue are precisely the same as those which hinder growth in fretting fatigue.

This book is laid out in such a way as to present as much underlying mechanics as possible. To this end, in the early chapters geometrically idealized contacts are discussed, partly so that closed form solutions to the salient parameters may be found, and partly because such geometries are an ideal vehicle for testing hypotheses of fretting fatigue experimentally, as will be discussed in Chapter 7. Thus, the first step in understanding any fretting problem is to understand the size and shape of contact, and the traction distribution itself. These topics are treated in Chapter 2 for both plane (two-dimensional) and axi-symmetric problems. Techniques for the solution of the problems, together with a wide range of closed-form answers are provided, including not only the surface traction but also the interior stress and displacement fields. The limitations of the techniques described are explained so that caution may be exercised in their application. The second phase in the solution is to quantify the surface damage; this may be done by finding the combination of stick and slip zones existing under partial slip conditions, together with the shear traction distribution, which may be quite different from that anticipated in a sliding contact; this material is covered in Chapter 3. More advanced topics in contact mechanics are covered in the next chapter, permitting the repertoire of solutions to be extended to those in strips, those found in contacts between dissimilar bodies and those arising from twisting. Also, numerical procedures for solving contact problems under more general conditions are discussed. All of the contact problems solved to this point are for geometrically smooth bodies. In crack initiation it is known that the surface finish plays a key rôle in defining the incubation period of a crack. In Chapter 5, therefore, we present a discussion of surface roughness, including both its numerical characterisation and its influence on local contact stresses. Also, the ideas of friction are developed, permitting a more complete understanding of the shear traction distribution.

In Chapter 6 we turn away from the contact itself to the study of any cracks formed. We present methods for the rigorous determination of crack tip stress intensity factors, so that fracture mechanics and contact mechanics may be combined to give a good description of the growth of fretting fatigue cracks. From there we go on to describe a range of possible fretting fatigue tests, highlighting the advantages and limitations of each one. Notes are provided on the analysis of the corresponding contact problems. The later chapters of the book describe the results of fretting fatigue tests, their interpretation, and their use in formulating design criteria. Crack initiation and propagation are treated separately, and in conclusion we summarise the present state of knowledge concerning possible criteria for the avoidance of fretting fatigue.

Chapter 2

Basic Contact Mechanics

2.1 Introduction

Many real fretting problems involve the contact of components of great complexity, such as the flanged joints depicted in the pipework photograph, fig. 2.1. In cases such as this it is really quite impossible to consider modelling the contact by anything other than a numerical method, and the technique chosen will in many cases be the finite element method. It must be remembered when doing such modelling that care needs to be exercised: not only must a carefully refined mesh in the neighbourhood of the contact be incorporated, but many other aspects need to be carefully studied, too. In the flanged joint, for example, what is the pre-load provided in the bolts? Are the contacting bodies elastically similar? How is the size of the contact patch determined - and is it known *a priori*? Which components adhere to each other, i.e. experience no relative displacement at any point, and which experience frictional effects? What is the history of loading experienced, i.e. do normal and tangential loads arise at the same time, or is there a phase shift between the maximum normal and tangential components of loading? These questions are often difficult to answer, but we hope that a consideration of the idealized problems presented in this and the next chapter will provide the analyst with physical insight. This will facilitate the choice of the correct boundary conditions for the investigation of complex real problems.

When the finite element problem described above has been run successfully, convergence has occurred and all the appropriate inequalities satisfied, the output obtained will give the following: the complete state of stress in the neighbourhood of the contact, the traction distribution in the surface, and the relative tangential displacement of contacting points within any slip region. These are the quantities which, from the mechanics perspective, must control the initiation, development and growth of fretting cracks. Additionally, there may be some bulk geometry dependence (for example if fretting debris is trapped), and microscopic geometry dependence. The latter will be associated with surface finish and influence the number of load reversals experienced by surface particles on an asperity scale (see §9.4).

Figure 2.1: Typical pipework arrangement showing a number of bolted flange joints.

The object of the analysis will be to predict conditions under which fretting cracks arise, and then to predict their rate of growth. It would be possible to carry out tests on actual geometries such as those depicted in fig. 2.1, but this would necessitate numerical modelling of the contact, usually by the finite element method. This will introduce some inaccuracy into the solution, particularly in the location of stick-slip zone boundaries, which are notoriously difficult to locate. It is therefore preferable to use tests employing well defined idealised geometries, so that the nature of the contact and the induced stresses/displacements are well-defined, easily controlled, repeatable, and insensitive to manufacturing imperfections. The experiments described in Chapter 7 are designed to achieve these objectives. The contacts to be described here have been studied by a number of authors and readers requiring treatment in greater depth may find Gladwell (1980) a useful reference.

A useful first step in analysing contacts is to carry out a classification; that which we introduce here is itself capable of giving some information about the nature of the contact, fig. 2.2. First, consider a cylinder, lightly pressed onto an elastic half-plane,

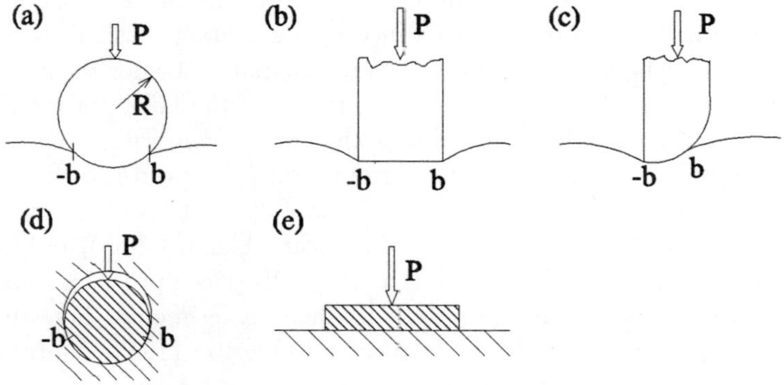

Figure 2.2: Characterisation of contacts; (a) Incomplete and non-conformal, (b) complete, (c) incomplete but with singularity, (d) incomplete and conformal, (e) receding.

fig. 2.2(a). As the two bodies are brought together contact is made along a line, and then, as the load is increased, this line will broaden to form a narrow strip. This is an example of an **incomplete** contact, i.e. its extent is not fixed geometrically, but is dependent on the applied load. Further, the contacting bodies have a common tangent at the edge of the contact, and the slope of the deformed surfaces is continuous there. It therefore seems physically reasonable (and this may be proved rigorously, (Hills, Nowell and Sackfield, 1993)) that the contact pressure also falls continuously to zero. In incomplete contacts the contact pressure distribution is only locally disturbed by minor imperfections in surface finish. By contrast, fig. 2.2(b) depicts a rigid, flat-ended punch pressed into an elastic plane. Here the size of the contact is independent of the load: it is therefore said to be **complete** contact. The two bodies do not have a common tangent at the edge of the contact, indeed the slope of the surface of the elastic half-plane is not continuous there, and the corresponding contact pressure is singular. A minor manufacturing flaw, for example a tiny burr near one end, may grossly influence the contact pressure distribution. It is possible to postulate contacts where a combination of these two characteristics arise, such as the D-shaped punch of fig. 2.2(c): in a case like this the contact pressure will be singular at the abrupt edge but fall continuously to zero at the other end of the contact; The extent of the contact will be dependent on the load.

A second kind of classification concerns *conformality.* Consider fig. 2.2(a) again. Under a light load the contact half-width, a, will be much less than the characteristic radius of the cylinder, R. Under these circumstances, i.e. providing $a << R$ the deformation of the cylinder will be confined to a tiny arc of its circumference. The cylinder may therefore be approximated by a half-plane when evaluating the deformation and stresses. This simplification is possible because the bodies do not conform, i.e. the largest characteristic contact dimension (a in this instance) is much less than the smallest characteristic radius of curvature of the contacting bodies (R in this instance). Figure 2.2(d) shows another incomplete contact, formed by pressing a cylinder radially into a hole of only slightly larger diameter. Under these conditions the arc of contact is a significant fraction of the radius of the hole, and neither body may be approximated by a half-plane. This is known as a conformal contact, and in order to solve the problem we must use a formulation appropriate to a disk and an infinite plane containing a hole, which is a much more difficult problem, though solutions have been provided by Persson (1964), Mostofi and Gohar (1980), and Goodman and Keer (1965). It is interesting to note that the half-space approximation is valid in practice up to quite large values of a/R; in their experimental investigation of contacting bodies using photoelasticity, Fessler and Ollerton (1957) quote an upper limit of 0.3, at which point the degree of conformality is quite severe.

In the flat-ended punch problem of fig. 2.2(b), it may be seen that for the punch itself approximation by a half-plane is inappropriate. Either a numerical solution using a formulation for a rectangle may be employed (Khadem and O'Connor, 1969), if the two bodies are of comparable stiffness, or, wherever possible, a solution may be employed in which the punch is assumed to be rigid. An unusual kind of contact is shown in fig. 2.2(e). Here, a thin, flat elastic plate is resting on an extended elastic block. Application of a central normal force will cause the outer edges to lift and the contact patch to **recede**. This class of contacts has been studied by Keer, Dundurs and Tsai (1972), Dundurs (1975) and Gladwell (1976). In this book we will concentrate our attention on contacts which are capable of being modelled using half-planes (or half-spaces); we can then analyse the influence of frictional forces without resorting to purely numerical techniques.

It is now appropriate to examine the origin and distribution of surface shearing tractions, and to assess, in turn, their influence on the contact pressure distribution. The general principles embodied in the results to be deduced are equally applicable to plane (two dimensional), axi-symmetric, or indeed arbitrarily shaped contacts. Suppose two elastically similar bodies are brought into normal contact, fig. 2.3. As this is done a mutual contact pressure develops, and surface particles within each body are compressed but also suffer a displacement parallel to the free surface, as will be shown. However, as the bodies are elastically similar the displacement of two corresponding particles, one on each body, will be the same, and there will be no tendency towards relative slip. This means that no frictional tractions arise and the solution to the problem is the same, whether the surfaces adhere or are well lubricated. If now a tangential force sufficient to cause sliding[1] is applied, shearing tractions arise which are everywhere limited by the coefficient of friction, f, i.e.

$$|q(x,y)| = -fp(x,y) \tag{2.1}$$

where q is the shear stress and p the direct stress (contact pressure, < 0). The

[1] By sliding we mean that there is macroscopic relative tangential motion between the two contacting bodies.

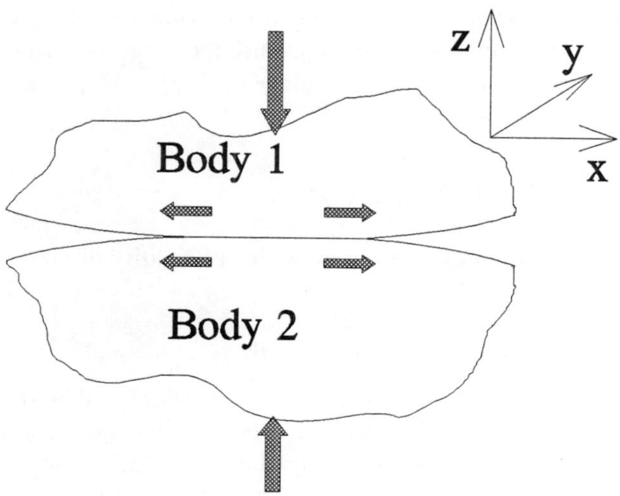

Figure 2.3: Normal contact between two bodies.

presence of the shearing tractions will cause a normal displacement of each of the contacting surfaces. However, because the shear tractions act in opposite directions over the surface of the two bodies, and because the two bodies are elastically similar, surface points on the two bodies move in the normal direction by the *same amount* so that the surface profiles are not modified, and hence the contact pressure distribution remains unchanged. In fretting contacts the applied shearing force, Q, is usually too small to cause sliding, i.e.

$$|Q| < fP \tag{2.2}$$

where P is the normal force. In this case the contact area consists of a mixture of stick zones, where particles adhere, and slip zones, where relative tangential motion occurs and the shearing traction is again limited by friction. The distribution of stick and slip zones is governed by the following considerations: in a slip zone the shear traction is given by equation (2.1), and the *direction* of the shear traction is dictated by the requirement that it opposes the direction of the increment of relative surface tangential displacement induced by the last load increment. Within the stick zones the shearing traction must be less than the limiting value, i.e.

$$|q(x,y)| < -fp(x,y) \tag{2.3}$$

Again, it will be noted that the shearing tractions on each body are equal in magnitude but opposite in direction, so that the normal surface displacements induced in each body are the same. There is therefore no difference in relative curvature and the contact pressure is unaffected. Hence, in all of the above problems - indentation, sliding, partial slip - the effects of normal and shear loading are entirely separate and may therefore be analyzed separately. The problem is uncoupled.

Suppose now that the two bodies under consideration, fig. 2.3, have different elastic constants. Now, when the two elements are brought together, the surface particles on each component suffer a tangential displacement, δ_i^T, which is proportional to the compliance of that component, i.e.

$$\delta_i^T(r,s) \propto \frac{1-\nu_i}{\mu_i} p(x,y) \qquad i = 1,2 \tag{2.4}$$

where ν_i is Poisson's ratio and μ_i is the modulus of rigidity of body i. Thus unless

$$\frac{1-\nu_1}{\mu_1} = \frac{1-\nu_2}{\mu_2}, \tag{2.5}$$

a relative surface tangential displacement will occur, which will be resisted by friction. In some regions this will be insufficient to prevent relative motion and slip will occur, whilst in other regions particles remain adhered to each other. The physical arguments which enable us to decide the stick and slip zone distributions are exactly the same as those put forward for the influence of a shearing force less than the sliding value. The problem is, however, even more complex as the surface shearing tractions cause a normal displacement of the surface, δ_i^N, i.e.

$$\delta_i^N(r,s) \propto \frac{1-\nu_i}{\mu_i} q(x,y) \qquad i = 1,2 \tag{2.6}$$

Again, unless equation (2.5) holds, there will be a change in the relative normal surface displacement, and hence a change in the relative surface profile, which will, in turn, affect the contact pressure distribution. The problem is 'coupled'. Similarly, in a sliding contact the surface profiles become changed from the unloaded ones, and hence the contact pressure is also different. Lastly, under tangential loading less than the limiting value a mixture of stick and slip zones arises which is different from the 'uncoupled' case. Further consideration of these cases is deferred until Chapter 4.

The last phase of classification is the simplest: whether the geometry is plane (two-dimensional), axi-symmetric, or generally three-dimensional. The last category admits of an analytical solution only when the contact is elliptical, and in any case requires a potential method (Hills, Nowell and Sackfield, 1993). We will therefore concentrate on plane and axi-symmetric problems.

2.2 Plane problems - formulation

Solution of the problem proceeds in two phases: first the contact problem itself may be solved to determine the contact pressure distribution; the internal stress, strain and displacement field is then found using the previously deduced surface tractions. The first step is therefore to produce a technique for finding the traction distribution, and the method employed is to establish and solve an integral equation.

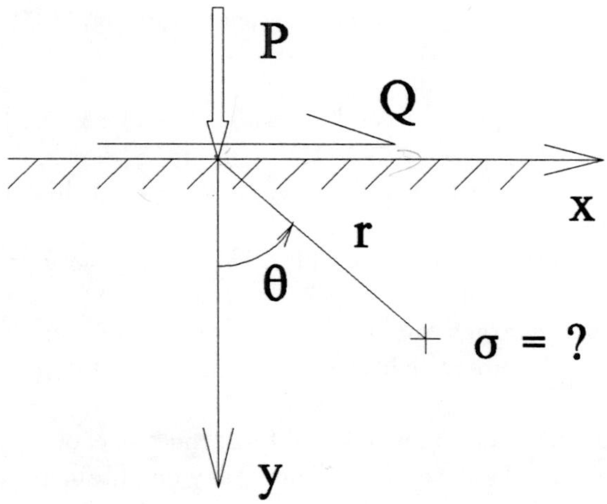

Figure 2.4: A half-plane subject to line normal and tangential forces.

Figure 2.4 shows a half-plane subject to a line force having normal and tangential components P, Q per unit length respectively. It is assumed that the half-plane is in a state of plane deformation, so that the state of stress may be given as an Airy stress function solution of the biharmonic equation (Timoshenko and Goodier, 1951). In this case this is

$$\phi(r,\theta) = -\frac{r\theta}{\pi}(P\sin\theta + Q\cos\theta), \tag{2.7}$$

and substitution into the biharmonic equation gives the stress field

$$\begin{aligned} \sigma_{rr} &= -\frac{2}{\pi r}(P\cos\theta - Q\sin\theta) \\ \sigma_{rr} &= \tau_{r\theta} = 0 \end{aligned} \tag{2.8}$$

The corresponding strain field may be found by substituting equation (2.8) into Hooke's Law, which for plane conditions is

$$\begin{aligned} \epsilon_{rr} &= \tfrac{1}{8\mu}\{\sigma_{rr}(1+\kappa) + \sigma_{\theta\theta}(\kappa-3)\} \\ \epsilon_{\theta\theta} &= \tfrac{1}{8\mu}\{\sigma_{\theta\theta}(1+\kappa) + \sigma_{rr}(\kappa-3)\} \\ \gamma_{r\theta} &= \tau_{r\theta}/\mu \end{aligned} \tag{2.9}$$

where $\kappa = 3 - 4\nu$ in plane strain, $\kappa = (3-\nu)/(1+\nu)$ in plane stress, μ is the modulus of rigidity and ν is Poisson's ratio. When once the strain field is known the displacements may be found by integration with the following outcome

$$2\mu u_r = -\frac{P}{2\pi}\{(\kappa-1)\theta\sin\theta - \cos\theta + (\kappa+1)\ln r\cos\theta\}$$

$$-\frac{Q}{2\pi}\{(\kappa-1)\theta\cos\theta+\sin\theta-(\kappa+1)\ln r\sin\theta\}+C_1 \quad (2.10)$$

$$\begin{aligned}2\mu u_\theta &= \frac{P}{2\pi}\{(\kappa-1)\theta\cos\theta-\sin\theta-(\kappa+1)\ln r\sin\theta\}\\ &+\frac{Q}{2\pi}\{-(\kappa-1)\theta\sin\theta-\cos\theta-(\kappa+1)\ln r\cos\theta\}+C_2\end{aligned} \quad (2.11)$$

where C_1, C_2 are arbitrary constants whose values may be found by specifying the displacements relative to some remote datum. It should be noted that the deeper the datum depth is set, the larger will be the value of the displacements. This inability to find the *absolute* displacement field is characteristic of all two-dimensional elasticity problems.

To find an influence function which will form the kernel of an integral equation we need to find the surface displacements. These may be obtained from equations (2.10) and (2.11) by setting $\theta = \pm\pi/2$, and converting to Cartesian coordinates. This gives

$$u = -P\left(\frac{\kappa-1}{8\mu}\right)\mathrm{sgn}(x)+Q\left(\frac{\kappa+1}{4\pi\mu}\right)\ln|x|+\frac{C_1}{2\mu} \quad (2.12)$$

$$v = -P\left(\frac{\kappa+1}{4\pi\mu}\right)\ln|x|-Q\left(\frac{\kappa-1}{8\mu}\right)\mathrm{sgn}(x)+\frac{C_2}{2\mu} \quad (2.13)$$

In principle, equations (2.12) and (2.13) could be used as the basis of a solution for contact problems, by determining the traction distribution required to produce a specified normal, and where appropriate, tangential, displacement. But the presence of the arbitrary constants C_1, C_2 is inconvenient, and it is normally preferable to work in term of the derivatives of $u(x), v(x)$ along the surfaces. Carrying out the differentiation and replacing the line forces by distributed tractions we find

$$\frac{\partial u}{\partial x} = \frac{\kappa-1}{4\mu}p(x)+\frac{\kappa+1}{4\pi\mu}\int\frac{q(\xi)d\xi}{x-\xi} \quad (2.14)$$

$$\frac{\partial v}{\partial x} = -\frac{\kappa-1}{4\mu}q(x)+\frac{\kappa+1}{4\pi\mu}\int\frac{p(\xi)d\xi}{x-\xi} \quad (2.15)$$

Equations (2.14) and (2.15) give the derivatives of the surface displacements at any point due to distributed direct and shear tractions $(p(\xi), q(\xi))$, applied over the surface. Suppose now that we have two bodies which are to be brought into contact by the application of normal and shearing forces P, Q, fig. 2.5. Two corresponding points on the surface will now have a relative displacement component in the y-direction, given by $v_1(x) - v_2(x)$. Employing equation (2.15) for each body, with an appropriate change of sign for body 1, we find that, in the contact patch

$$\frac{1}{A}\frac{\partial h}{\partial x} = \frac{1}{\pi}\int\frac{p(\xi)d\xi}{x-\xi}-\beta q(x) \quad (2.16)$$

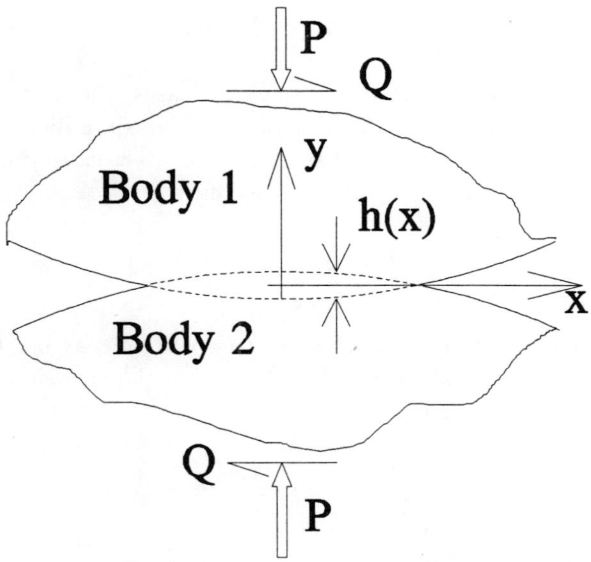

Figure 2.5: Line contact between two elastically deformable bodies subject to a normal force, P, and a tangential force, Q.

where $h(x) = v_1(x) - v_2(x)$ which is the amount of overlap which would occur if the contacting bodies could freely interpenetrate each other and

$$A = \frac{\kappa_1 + 1}{4\mu_1} + \frac{\kappa_2 + 1}{4\mu_2} \quad \text{- The composite compliance} \tag{2.17}$$

$$\beta = \frac{\Gamma(\kappa_1 - 1) - (\kappa_2 - 1)}{\Gamma(\kappa_1 + 1) + (\kappa_2 + 1)} \quad \text{- Dundurs' parameter (Dundurs, 1969)} \tag{2.18}$$

$$\Gamma = \mu_2/\mu_1$$

Normally, contact problems exist under conditions of transverse plane strain, when these quantities take on the form

$$A = \frac{1 - \nu_1}{\mu_1} + \frac{1 - \nu_2}{\mu_2} \equiv 2\left\{\frac{1 - \nu_1^2}{E_1} + \frac{1 - \nu_2^2}{E_2}\right\} \tag{2.19}$$

$$\beta = \frac{1}{2A}\left\{\frac{1 - 2\nu_1}{\mu_1} - \frac{1 - 2\nu_2}{\mu_2}\right\} \equiv \frac{1}{A}\left\{\frac{(1 - 2\nu_1)(1 + \nu_1)}{E_1} - \frac{(1 - 2\nu_2)(1 + \nu_2)}{E_2}\right\} \tag{2.20}$$

where E_i is the Young's modulus of body i. Similarly, if the relative tangential displacement of corresponding points in contact is $g(x)$, equation (2.15) yields the following expression

$$\frac{1}{A}\frac{\partial g}{\partial x} = \frac{1}{\pi}\int \frac{q(\xi)d\xi}{x - \xi} + \beta p(x) \tag{2.21}$$

Equation (2.16) and, where appropriate, (2.21), may be used to solve a very wide range of plane contact problems.

Let us, for the time being, restrict ourselves to cases where the second term on the right hand side of equation (2.16) disappears. This simplification applies when no shear tractions arise, i.e. either because that the contact is frictionless, or because β (equation 2.18) is zero. The equation then becomes

$$\frac{1}{A}\frac{\partial h}{\partial x} = \frac{1}{\pi}\int_{-a}^{a}\frac{p(\xi)d\xi}{x-\xi} \qquad |x| \leq a \tag{2.22}$$

where the contact patch extends from $x = -a$ to $x = a$. It may be shown (Muskhelishvili, (1953), Hills, Nowell and Sackfield, (1993)), that the inversion of this equation, that is, the solution for $p(x)$, given $h(x)$, is

$$p(x) = -\frac{w(x)}{A\pi}\int_{-a}^{a}\frac{h'(\xi)d\xi}{w(\xi)(\xi-x)} + Cw(x) \tag{2.23}$$

where $h'(x) \equiv \partial h/\partial x$ and the fundamental function, $w(x)$, can take one of four forms, dependent upon the required behaviour of $p(x)$ at the end points (see Table 2.1). Examples of the use of this technique are given in §2.3.

Behaviour at end		$w(x)$	C
$x=-a$	$x=a$		
N	S	$\sqrt{(a+x)/(a-x)}$	0
S	N	$\sqrt{(a-x)/(a+x)}$	0
S	S	$1/\sqrt{a^2-x^2}$	Non Zero
N	N	$\sqrt{a^2-x^2}$	0*

S = Singular, N = Non-singular

* The additional consistency condition $\int_{-a}^{a}\frac{h'(\xi)}{w(\xi)}d\xi = 0$ must be imposed.

Table 2.1: Fundamental functions for use in inversion of singular integral equations of the first kind (equation 2.22).

In fretting experiments we shall also need to understand details of the internal stress field and the surface displacements. For plane problems the most efficient way of deducing these quantities is via the Muskhelishvili potential (Muskhelishvili, (1953), Gladwell (1980), Hills, Nowell, and Sackfield, (1993)).

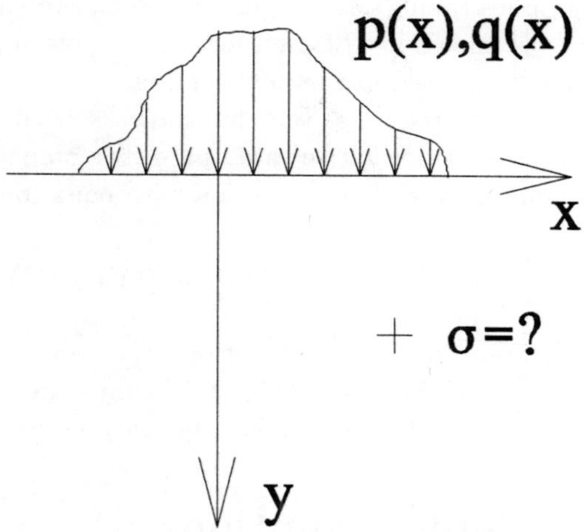

Figure 2.6: A half plane loaded by normal and shear tractions, $p(x)$, $q(x)$, respectively, applied over the contact patch.

The potential, ϕ, is a function of position, z, which is a complex coordinate ($= x + iy$). The geometry and co-ordinate set used are shown in figure 2.6, and the potential itself is found by carrying out the following contour integral along the line of the contact:

$$\phi(z) = \frac{1}{2\pi i} \int_{\text{contact}} \frac{p(t) - iq(t)}{t - z} dt \tag{2.24}$$

Here, $p(t), q(t)$ are arbitrary traction distributions; for sliding contacts these variables are related by equation (2.1), and hence

$$\phi(z) = \frac{1 - if}{2\pi i} \int_{\text{contact}} \frac{p(t)dt}{t - z} \tag{2.25}$$

When once the potential is found the stress components are defined by

$$\sigma_{xx} + \sigma_{yy} = 2(\phi(z) + \bar{\phi}(\bar{z})) \tag{2.26}$$

$$\sigma_{yy} - \sigma_{xx} + 2i\tau_{xy} = 2((\bar{z} - z)\phi'(z) - \bar{\phi}(z) - \phi(z)) \tag{2.27}$$

It will be seen that further operations on the potential ϕ are required: $\phi'(z)$ implies differentiation with respect to $z, \bar{\phi}(z)$ implies taking the conjugate of the function which, for potentials defined by equation (2.25), may be found directly from

$$\bar{\phi}(z) = -\frac{1 + if}{1 - if}\phi(z) \tag{2.28}$$

and the notation $\bar{\phi}(\bar{z})$ implies taking the conjugate of ϕ and replacing the argument by $\bar{z}$. Hence, when once ϕ is known it is straightforward to obtain the other functions needed. These formulae will be used in the next section.

Displacement fields are also required, but, for the reason cited when discussing equations (2.10, 2.11) only their derivatives are unique for plane problems. By employing Hooke's Law (equation 2.9) in conjunction with equations (2.26), (2.27) we arrive at

$$2\mu\left(\frac{\partial u}{\partial x}+i\frac{\partial v}{\partial y}\right)=(\bar{z}-z)\bar{\phi}'(\bar{z})+\phi(\bar{z})+\kappa\phi(z) \tag{2.29}$$

The only further complication here is the function $\bar{\phi}'(\bar{z})$ which implies taking the function ϕ, differentiating with respect to z, taking the conjugate and finally replacing the argument z by $\bar{z}$. In practice these operations produce no difficulties.

2.3 Plane problems - solutions

In this section specific geometries will be considered and solutions found first for the contact pressure and subsequently for the field quantities, using the techniques described in the last section. Here, only uncoupled problems will be considered (the influence of material mismatch is treated in Chapter 4).

2.3.1 Contact of cylinders - the Hertz problem

The classic well-defined contact problem admitting of an exact solution is that occurring between two parallel, infinitely long cylinders. In practice the ideal problem does not arise because of end effects which may cause local severe stress concentrations; and for this reason the rollers in bearings are very often barrelled. But well away from the ends the state of stress occurring in these cases is nearly two-dimensional, so that the plane solution is a satisfactory approximation.

The solution described applies if both bodies are curved to a circular profile (concave or convex) providing that the radii of curvature are large in comparison with the contact width, to achieve non-conformality (see §2.1). This is necessary so that both elements may be approximated by half-planes, but it has the further implication that the local relative profile at the point of contact may be idealised as a parabola, which permits $h'(x)$ in equation (2.23) to be written in a simple form, and the resulting integral to be evaluated analytically. The equation giving the amount of overlap in freely interpenetrating bodies (fig. 2.5) is

$$h(x)=\Delta-\frac{1}{2}kx^2 \tag{2.30}$$

where k is the relative curvature, given by

$$k=\frac{1}{R_1}+\frac{1}{R_2}$$

and R_1, R_2 are the radii of the contacting surfaces. Note that if one of the bodies is concave the corresponding radius of curvature must be taken as negative. For the reason discussed in the previous section we shall work not in terms of the profile, but its slope, so that

$$\frac{dh}{dx} = -kx \tag{2.31}$$

This result may now be substituted into equation (2.23). It is anticipated that the contact pressure will be zero at both ends of the contact and so from Table 2.1 it may be seen that $C = 0$ and the appropriate weight function is

$$w(x) = \sqrt{a^2 - x^2} \tag{2.32}$$

Equation (2.23) therefore becomes

$$p(x) = -\frac{\sqrt{a^2 - x^2}}{A\pi} \int_{-a}^{a} \frac{k\xi d\xi}{\sqrt{a^2 - \xi^2}(\xi - x)}. \tag{2.33}$$

Evaluation of the integral then gives

$$p(x) = -\frac{k}{A}\sqrt{a^2 - x^2} \tag{2.34}$$

The resulting pressure distribution is clearly elliptical but the result is of limited use in this form as the contact half-width, a, is unknown. This may be remedied by ensuring equilibrium between the contact pressure and the applied load, P, i.e.

$$\begin{aligned} P &= -\int_{-a}^{a} p(\xi) d\xi \\ &= \frac{\pi k a^2}{2A} \end{aligned} \tag{2.35}$$

Perhaps the most convenient way to express these results is in the following form:

$$a^2 = \frac{2PA}{\pi k} \tag{2.36}$$

and

$$p(x) = -p_0\sqrt{1 - (x/a)^2} \tag{2.37}$$

where

$$p_0 = \frac{2P}{\pi a} \tag{2.38}$$

is the peak contact pressure. With the traction distribution solved, we may now find the stress and displacement fields. First, the Muskhelishvili potential is found by substituting equation (2.37) into equation (2.25), giving

$$\phi(z) = \frac{-(1 - if)}{2\pi i}\frac{p_0}{a} \int_{-a}^{a} \frac{\sqrt{a^2 - x^2}dx}{x - z} \tag{2.39}$$

$$\phi(z) = \frac{p_0}{2a}(i+f)\left(z - \sqrt{z^2 - a^2}\right) \tag{2.40}$$

Differentiation of this result gives

$$\phi'(z) = -\frac{p_0}{2a}(i+f)\left(1 - \frac{z}{\sqrt{z^2 - a^2}}\right) \tag{2.41}$$

whilst conjugation gives

$$\bar{\phi}(z) = -\frac{p_0}{2a}(-i+f)\left(z - \sqrt{z^2 - a^2}\right) \tag{2.42}$$

and lastly $\bar{\phi}(\bar{z})$ is given by

$$\bar{\phi}(\bar{z}) = -\frac{p_0}{2a}(-i+f)\left(\bar{z} - \sqrt{\bar{z}^2 - a^2}\right) \tag{2.43}$$

From equations (2.40) to (2.43) the state of stress and displacement (to within an arbitrary constant) may be found from equations (2.26, 2.27, 2.29). Explicit values may be readily found by using complex arithmetic in a computer code. However, occasionally it may be preferred to work in real arithmetic with the following result (see also Smith and Liu, 1953, Radzimovsky, 1953, Poritsky, 1950 and M'Ewan, 1949):

$$\begin{aligned}
\frac{\sigma^n_{xx}}{p_0} &= \frac{y}{a}\left\{2 - \frac{s}{\sqrt{a^2+s^2}} - \frac{\sqrt{a^2+s^2}}{s} - \frac{x^2 s^3 a^2}{(a^2+s^2)^{\frac{3}{2}}(s^4+y^2a^2)}\right\} \\
\frac{\sigma^n_{zz}}{p_0} &= \frac{-2\nu y}{a}\left\{\frac{\sqrt{a^2+s^2}}{s} - 1\right\} \\
\tau^n_{xz} &= \tau^n_{yz} = 0 \\
\frac{\sigma^n_{yy}}{p_0} &= \frac{-y^3 a\sqrt{a^2+s^2}}{s(s^4+y^2a^2)} \\
\frac{\tau^n_{xy}}{p_0} &= -\frac{xy^2 sa}{\sqrt{a^2+s^2}(s^4+y^2a^2)} \\
\frac{\sigma^t_{xx}}{fp_0} &= \frac{-2x}{a}\left(1 - \frac{s}{\sqrt{a^2+s^2}}\right) + \frac{xy^2 sa}{\sqrt{a^2+s^2}(s^4+y^2a^2)} \\
\frac{\sigma^t_{zz}}{fp_0} &= \frac{-2\nu x}{a}\left(1 - \frac{s}{\sqrt{a^2+s^2}}\right) \\
\frac{\sigma^t_{yy}}{f} &= \tau^n_{xy} \\
\frac{\tau^t_{xy}}{f} &= \sigma^n_{xx} \\
\tau^t_{xz} &= \tau^t_{yz} = 0
\end{aligned} \tag{2.44}$$

where the superscripts n and t denote stresses due to normal and tangential tractions respectively, ν is Poisson's ratio, and s^2 is the largest root of the quadratic equation

$$\frac{x^2}{a^2+s^2}+\frac{y^2}{s^2}=1 \tag{2.45}$$

i.e.

$$s^2=\frac{1}{2}\left\{-(a^2-x^2-y^2)+\sqrt{(a^2-x^2-y^2)^2+4y^2}\right\} \tag{2.46}$$

It should be noted that within the contact patch as $y \to 0$, $s \to 0$, so that the y/s remains finite, i.e.

$$\underset{y\to 0}{Lim}\left(\frac{y^2}{s^2}\right)=a^2-x^2 \qquad |\, x\,| < a \tag{2.47}$$

Exterior to the contact, as $y \to 0$, s remains finite, and hence

$$\underset{y\to 0}{Lim}\left(\frac{y}{s}\right)=0 \qquad |\, x\,| > a \tag{2.48}$$

Fretting causes surface damage, so that the surface stress and displacement are of particular importance. Within the contact patch the stresses at the surface are given by

$$\frac{\sigma^n_{xx}}{p_0}=\frac{\sigma^n_{yy}}{p_0}=\frac{\tau^t_{xy}}{fp_0}=-\frac{\sqrt{a^2-x^2}}{a} \tag{2.49}$$

$$\frac{\sigma^t_{xx}}{p_0}=-2\frac{x}{a} \tag{2.50}$$

$$\frac{\sigma^t_{yy}}{fp_0}=\frac{\tau^n_{xy}}{p_0}=0 \tag{2.51}$$

$$\frac{\sigma^n_{zz}}{p_0}=-2\nu\frac{\sqrt{a^2-x^2}}{a} \tag{2.52}$$

$$\frac{\sigma_{zz^t}}{fp_0}=-2\nu\frac{x}{a} \tag{2.53}$$

Exterior to the contact *all* components of stress at the surface due to normal pressure vanish, whilst those associated with tangential shearing traction are

$$\frac{\sigma^t_{yy}}{fp_0}=\frac{\tau^t_{xy}}{fp_0}=0 \tag{2.54}$$

$$\frac{\sigma^t_{xx}}{fp_0}=\frac{-2x}{a}+\frac{2\sqrt{x^2-a^2}}{a}\mathrm{sgn}(x) \tag{2.55}$$

$$\frac{\sigma^t_{zz}}{fp_0}=-2\nu\left[x+\frac{\sqrt{x^2-a^2}}{a}\mathrm{sgn}(x)\right] \tag{2.56}$$

The *relative* surface displacement for **normal** contact between cylinders is useful in calculating partial slip, and may be deduced from the stress field by using Hooke's law and integrating (Poritsky (1950), Hills, Nowell and Sackfield (1993)). The results quoted below are exact if the two bodies are elastically similar. They may also be used if the two bodies are elastically dissimilar, but it should be emphasised that they are obtained via an 'uncoupled' solution to equation (2.16), so that they are only approximately true under these circumstances. The normal displacement component is

$$v(x) = -\frac{AP}{\pi}\left[\left(\frac{x}{a}\right)^2 + C_0\right] \qquad |x| \leq a \tag{2.57}$$

$$v(x) = -\frac{AP}{\pi}\left[\ln|X| + \frac{1}{2X^2} + C_1\right] \qquad |x| \geq a \tag{2.58}$$

where

$$X = \left|\frac{x}{a}\right| + \sqrt{\left(\frac{x}{a}\right)^2 - 1}$$

and A is the composite compliance (equation 2.19). The relative surface tangential displacement is given by

$$u(x) = \frac{2A\beta P}{\pi}(\pi + C_2) \qquad x < -a \tag{2.59}$$

$$u(x) = \frac{2A\beta P}{\pi} C_2 \qquad x > a \tag{2.60}$$

$$u(x) = \frac{2A\beta P}{\pi}\left(\cos^{-1}\left(\frac{x}{a}\right) - \left(\frac{x}{a}\right)\sqrt{1-(x/a)^2}\right) + C_2 \qquad |x| \leq a \tag{2.61}$$

where β is Dundurs' constant (equation 2.20). The constants C_0, C_1, C_2 are arbitrary but C_0, C_1 should be chosen so that the displacements are continuous at $x = \pm a$, and if C_2 is taken to be $-\pi/2$ anti-symmetry of the tangential displacement is assured. The influence of **shearing** tractions may be found by exploiting the 'symmetry' of displacements apparent from equations (2.12) and (2.13) with the result that

$$u_t(x) = -f v_n(x) \tag{2.62}$$

$$v_t(x) = f u_n(x) \tag{2.63}$$

However, considerable caution should be exercised in using these equations, particularly (2.63) as strictly speaking the traction distribution will be perturbed from a semi-ellipse if $\beta \neq 0$.

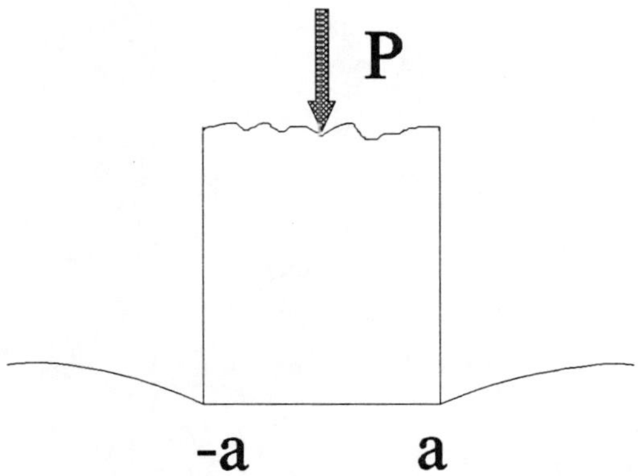

Figure 2.7: Indentaticn by a flat-ended punch.

2.3.2 Indentation by a rigid flat-ended punch

The classical complete contact of interest in fretting is that involving indentation by a 'punch' having a square end, fig. 2.7. It might be thought that the solution to be presented has relevance to the standard 'bridge' geometry fretting fatigue test, but it should be emphasised that this would only be so if the feet of the bridge were very stiff in comparison to the stiffness of the specimen, so that the latter accommodated all of the imposed displacement. This is because the deformation of the feet, which are of strictly finite width, may not be modelled as if they were half-planes; there is no material on either side to support the feet, and a formulation appropriate to a rectangular domain must be used, or a numerical technique.

However, although the solution for a rigid punch has its limitations, the fact that all the field quantities are known in closed form may be useful. As the end of the punch is flat, its gradient at all points is zero and hence

$$\frac{dh}{dx} = 0 \qquad \mid x \mid \leq a \tag{2.64}$$

As the contact is complete we expect the contact pressure to be singular at each end, and hence, from Table 2.1

$$w(x) = 1/\sqrt{a^2 - x^2} \tag{2.65}$$

Further, it will be seen that the integrand in equation (2.23) vanishes everywhere, and hence the solution of the integral equation itself is simply given by

$$p(x) = \frac{C}{\sqrt{a^2 - x^2}} \tag{2.66}$$

The constant, C, may be computed by invoking equilibrium with the external load,

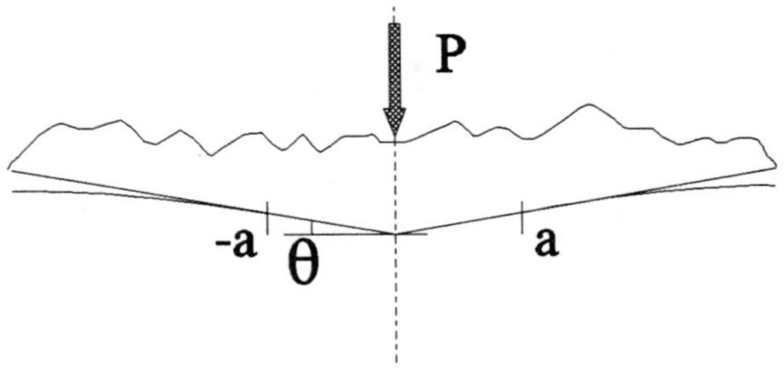

Figure 2.8: Indentation by a shallow wedge.

equation (as in 2.35), giving

$$p(x) = -\frac{P}{\pi\sqrt{a^2 - x^2}} \tag{2.67}$$

The next step in the solution is to determine the Muskhelishvili potential, using equation (2.25). As before, for generality we will assume that the punch may be sliding, but it must be borne in mind that in this case the solution will be strictly correct only if Dundurs' constant, β, in equation (2.20), vanishes, and hence that the indented body is incompressible, i.e. $\nu = \frac{1}{2}$ [2]. Evaluation of equation (2.25) gives

$$\phi(z) = \frac{(i + f)P}{2\pi\sqrt{z^2 - a^2}} \tag{2.68}$$

and

$$\phi'(z) = \frac{-(i + f)Pz}{2\pi(z^2 - a^2)^{\frac{3}{2}}} \tag{2.69}$$

The other functions may be derived using the operations described in the previous section, and given for the case of contact of cylinders above.

2.3.3 Indentation by a wedge

Indentation by a shallow wedge, fig. 2.8, provides another interesting example of incomplete contact. As we wish to retain the assumption that both bodies may have a finite compliance, each must be capable of being modelled as a half-plane, which means that the angle θ must be small. With this restriction it will be seen that the gradient of the profile of the relative normal displacement is given by

$$\frac{dh}{dx} = -\theta \operatorname{sgn}(x) \tag{2.70}$$

[2]Since the punch is rigid, β can only be zero if the indented body has a Poisson's ratio of 0.5 (see equation 2.18).

and hence the integral equation to be inverted is given by

$$-\frac{\theta}{A}\text{sgn}(x) = \frac{1}{\pi}\int_{-a}^{a}\frac{p(\xi)d\xi}{x-\xi} \qquad |x| \leq a \tag{2.71}$$

Taking the 'bounded both ends' solution from table 2.1 we find

$$p(x) = -\frac{\sqrt{a^2-x^2}}{A\pi}\theta\int_{-a}^{a}\frac{\text{sgn}(\xi)d\xi}{\sqrt{a^2-\xi^2}(\xi-x)} \tag{2.72}$$

i.e.

$$p(x) = -\frac{2\theta}{\pi A}\cosh^{-1}\left(\frac{a}{|x|}\right) \tag{2.73}$$

Once again the contact patch size may be found by ensuring equilibrium with the external load is satisfied, (cf. 2.35) giving

$$a = \frac{PA}{2\theta} \tag{2.74}$$

and hence

$$p(x) = -\frac{P}{\pi A}\cosh^{-1}\left(\frac{a}{|x|}\right)$$

or

$$p(x) = -\frac{2\theta}{\pi A}\cosh^{-1}\left(\frac{PA}{2\theta|x|}\right) \tag{2.75}$$

2.3.4 Indentation by a punch producing constant pressure

This problem is of some practical interest as it might be anticipated that if a soft punch is rubbed over a much harder half-plane, wear of the punch will occur so as to produce a profile giving a constant pressure distribution. Care must be exercised in using the solution if both bodies are assumed to be elastic, to ensure that each may reasonably be approximated by a half-plane.

If the contact pressure is constant, and contact is made over a strip of width $2a$, the pressure is given by

$$p = P/2a \tag{2.76}$$

and hence equation (2.22) is now simply an integral rather than an integral equation. Evaluating the integral gives

$$\frac{1}{A}\frac{dh}{dx} = -\frac{P}{2a\pi}\ln\left(\frac{a+x}{a-x}\right) \qquad |x| \leq a \tag{2.77}$$

and hence

$$h(x) = \frac{PA}{2\pi a}\left\{(a+x)\ln\left(1+\frac{x}{a}\right)+(a-x)\ln\left(1-\frac{x}{a}\right)\right\}+C \tag{2.78}$$

The corresponding Muskhelishvili potential is

$$\phi(z) = -\frac{P}{4\pi a}(i+f)\ln\left(\frac{z-a}{z+a}\right) \tag{2.79}$$

and

$$\phi'(z) = \frac{P}{2\pi a}\frac{(i+f)}{\sqrt{z^2-a^2}} \tag{2.80}$$

Further geometries have been studied and are summarized in the books by Sneddon (1951), Gladwell (1980), Johnson (1985) and Hills, Nowell and Sackfield (1993).

2.4 Axi-symmetric problems - formulation

Plane problems are attractive to the analyst because it is possible to obtain closed form solutions. Axi-symmetric problems are more attractive to the experimentalist because of the lack of edge effects which induce severe states of stress and promote the initiation of cracks. For this reason the sphere-on-flat or 'barrelled roller' configuration has much to recommend it. The solutions to be presented here are strictly axi-symmetric, and it is again assumed that the contacting components may be represented as semi-infinite bodies. Further, although in fretting problems there will be shearing forces present, provided that the bodies are elastically similar their presence will cause no change in the bodies' relative curvature; the solutions will be exact. They will also be exact for elastically dissimilar bodies provided that the contact is perfectly lubricated, i.e. no shear tractions arise, and we shall therefore represent the composite compliance of the materials by A, equation (2.19).

Derivation of the governing integral equation is a lengthy procedure, and is treated by Green and Zerna (1954), and Collins (1959). Gladwell (1980) or Hills, Nowell and Sackfield (1993) may be consulted for details. Here, we simply state that if two bodies are pressed together and produce an axi-symmetric contact, fig. 2.9 , the relationship between the contact pressure $p(r)$ and the degree of overlap of freely interpenetrating bodies, $h(r)$, is given by

$$h^*(r) = -A\int_r^a \frac{sp(s)ds}{\sqrt{s^2-r^2}} \tag{2.81}$$

where $h^*(r)$ is a transformed displacement, defined by

$$h^*(r) = \frac{d}{dr}\left\{\int_o^r \frac{sh(s)ds}{\sqrt{r^2-s^2}}\right\} \tag{2.82}$$

This two-stage process is necessary as a direct formulation using the physical surface profile $h(r)$ introduces a kernel composed of elliptic integrals, and its inversion is impossible. By contrast, when once $h^*(r)$ has been found, equation (2.81), which is of the Abel type, may be inverted directly (Collins (1959), Barber, (1983)) to give

$$p(r) = \frac{2}{\pi r A}\frac{d}{dr}\left\{\int_r^a \frac{sh^*(s)ds}{\sqrt{s^2-r^2}}\right\} \tag{2.83}$$

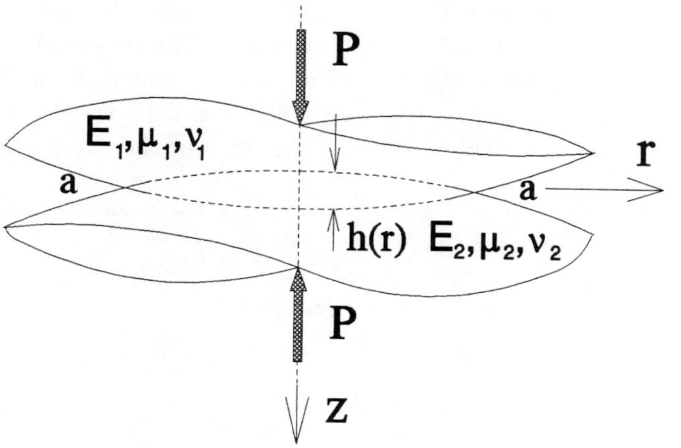

Figure 2.9: Contact between two axi-symmetric deformable bodies.

Thus, solution is a straightforward but two-stage process: $h(r)$ is substituted into equation (2.82) to determine $h^*(r)$ which is then inserted into equation (2.83) to determine the contact pressure distribution.

The second part of the solution is to determine the complete stress and displacement fields. Although partial slip contacts are to be expected in many fretting problems we shall concentrate here on finding the stresses induced by *sliding* contacts as the solution we shall obtain enables the effect of both direct and shear tractions to be found at the same time. The co-ordinate set used exploits the geometrical axi-symmetry of the problem; z is directed into the body and r is a radial co-ordinate, fig. 2.9. The stress field developed is expressed in Cartesian co-ordinates, where the x and y directions lie in the plane of the surface. We shall assume that the contact is sliding in the x-direction, and so the prescribed surface tractions are given by

$$\sigma_{zz}(r) = \frac{\tau_{zx}(r)}{f} = \begin{matrix} p(r) & r \leq a \\ 0 & r > a \end{matrix} \tag{2.84}$$

The starting point of the solution is the representation of the general displacement field for a half-space by Love (1927) in the form;

$$2\mu u = -(1-2\nu)N_x - zN_{xz} + 2\nu T_{xx} + 2T_{zz} - zT_{xxz} \tag{2.85}$$

$$2\mu v = -(1-2\nu)N_y - zN_{yz} + 2\nu T_{xy} - zT_{xyz} \tag{2.86}$$

$$2\mu w = 2(1-\nu)N_z - zN_{zzz} + (1-2\nu)T_{zx} - zT_{xzz} \tag{2.87}$$

where ν is Poisson's ratio, μ the modulus of rigidity, u, v, w are the displacements in the x, y, z directions respectively, and the potentials N and T will be associated with the normal and tangential components of loading respectively. Subscripts are variables of differentiation. To verify that these are acceptable representations of the solution to an elasticity problem we need to verify that they satisfy conditions of compatibility and equilibrium. Differentiation of equations (2.85 to 2.87) will yield the strains, and use of Hooke's Law then gives the stress field, which are the quantities required. The details of the algebra will not be given, but the stress field is, explicitly;

$$\sigma_{xx} = -N_{xx} - 2\nu N_{yy} - zN_{xxz} + 2(1+\nu)T_{xzz} + 2\nu T_{xxx} - zT_{xxxz} \tag{2.88}$$

$$\sigma_{yy} = -N_{yy} - 2\nu N_{xx} - zN_{yyz} - 2\nu T_{xxx} - zT_{xyyz} \tag{2.89}$$

$$\sigma_{zz} = N_{zz} - zN_{zzz} - zT_{xzzz} \tag{2.90}$$

$$\tau_{xy} = -(1-2\nu)N_{xy} - zN_{xyz} + 2\nu T_{xxy} + T_{yzz} - zT_{xxyz} \tag{2.91}$$

$$\tau_{yz} = -zN_{yzz} - zT_{xyzz} \tag{2.92}$$

$$\tau_{zx} = -zN_{xzz} + T_{zzz} - zT_{xxzz} \tag{2.93}$$

It will be seen that, on the surface

$$\sigma_{zz} \equiv N_{zz} \tag{2.94}$$

$$\tau_{zx} \equiv T_{zzz} \tag{2.95}$$

$$\tau_{yz} \equiv 0 \tag{2.96}$$

We need to establish potentials which satisfy equations (2.84) together with (2.94) and (2.95), and the most convenient way to do this is to use an integral representation. We choose

$$N = Im\left\{\int_0^a g(t)\left\{z_1 \ln(z_1 + R_1) - R_1\right\} dt\right\} \tag{2.97}$$

$$T = Im\left\{\int_0^a h(t)\left\{\frac{1}{2}\left(z_1^2 - \frac{1}{2}r^2\right)\ln\left(z_1 + R_1\right) - \frac{3}{4}z_1 R_1 + \frac{1}{4}r^2\right\} dt\right\} \tag{2.98}$$

where $g(t)$, $h(t)$ [3] are assumed real functions of t and

$$z_1 = \frac{z + it}{a}, \tag{2.99}$$

$$R_1^2 = \frac{x^2 + y^2 + (z+it)^2}{a^2} \tag{2.100}$$

The reason for this choice will become clear when the relation between the kernels $g(t)$, $h(t)$ and the tractions $p(r)$, $fp(r)$ is established. First, equation (2.97) is differentiated within the integral twice with respect to z, and equation (2.98) is differentiated three times. Combining with equations (2.84, 2.94, 2.95) we find

$$Im\left\{\int_0^a \frac{g(t)}{R_1} dt\right\} = \frac{1}{f} Im\left\{\int_0^a \frac{h(t)}{R_1} dt\right\} = \begin{matrix} p(r), & r \leq a \\ 0, & r > a \end{matrix} \tag{2.101}$$

[3]The quantities $g(t)$, $h(t)$ defined by these equations are not related in any way to the relative surface displacements $g(x)$, $h(x)$ employed earlier in this chapter.

The integrands have no imaginary part on $z = 0$, $r > a$ and we find the following Abel equations for the kernels $g(t)$, $h(t)$:

$$\int_r^a \frac{g(t)}{\sqrt{\{t^2 - r^2\}}} dt = \frac{1}{f} \int_r^a \frac{h(t)}{\sqrt{\{t^2 - r^2\}}} dt = -p(r) \qquad r \leq a \tag{2.102}$$

Their inversion (Barber, 1983) is

$$g(t) = \frac{h(t)}{f} = -\frac{2}{\pi} \frac{d}{dt} \left\{ \int_t^a \frac{rp(r)}{\sqrt{\{r^2 - t^2\}}} dr \right\} \tag{2.103}$$

When once the kernels are obtained for a particular problem it will be seen that the potentials themselves are implicitly known via equations (2.97, 2.98) and hence the derivatives needed for finding the stresses and displacements. In practice the resulting integrals are rather complex, and are given in Table 2.2. Also, although these results are complete and accurate for every point within the half-space, care is needed in their evaluation on the surface, and where possible the state of surface stress will be given explicitly.

2.5 Axi-symmetric problems - solutions

We will now use the results cited in the last section to find the surface tractions and stress field resulting from a wide range of axi-symmetric contacts. In this section the cylindrical co-ordinate set shown in fig. 2.9 will be retained for the first phase of each analysis (i.e. the determination of the surface tractions). Stress components will then be given in a Cartesian $Oxyz$ co-ordinate set, with the z-axis directed into the body.

2.5.1 Contact of spheres - the Hertz problem

As in the case of line contacts, the assumptions are made that no shear tractions arise, i.e. that the contact is either between elastically similar bodies or is perfectly lubricated, and that the contact radius is small in comparison with the radii of the contacting bodies. The latter restriction ensures that each may be replaced by a half-space and that the function $h(r)$ representing the total deformation to be accommodated in the bodies may be represented by a rotated parabola, i.e.

$$h(r) = \Delta - \frac{1}{2} kr^2 \tag{2.104}$$

where Δ is the approach of two remote points and k is the relative curvature, given by

$$k = \frac{1}{R_1} + \frac{1}{R_2} \tag{2.105}$$

and R_1, R_2 are the radii of the contacting spheres. Substituting into equation (2.82) we find the transformed displacement

$$h^*(r) = -(\Delta - kr^2) \tag{2.106}$$

and substituting now into equation (2.83) we find

$$p(r) = \frac{-2}{\pi r}\frac{d}{dr}\left\{\int_r^a \frac{s(\Delta - ks^2)}{\sqrt{s^2 - r^2}}\, ds\right\} \tag{2.107}$$

i.e.

$$p(r) = \frac{2}{\pi A}\left\{-\frac{\Delta}{\sqrt{a^2 - r^2}} - k\left[\frac{-r^2}{\sqrt{a^2 - r^2}} + \sqrt{a^2 - r^2}\right]\right\} \tag{2.108}$$

This interesting result expresses the most general physically acceptable solution for the contact pressure between two spheres. It gives explicitly a solution for $p(r)$ only if both Δ, the approach, and a, the radius of contact, are specified. It also includes a singular tension term, which varies as $(a^2 - r^2)^{-\frac{1}{2}}$, corresponding to the existence of cohesive or adhesive tensions. Physically, this is equivalent to a constant penetration (see §2.5.2). Analytically, the extra generality of the solution is because the formulation is cast in terms of displacements rather than displacement gradients; this, in turn, is because in non-planar geometries the absolute displacement of particles is known. In most engineering contacts the quality of the surface finish precludes the presence of adhesion, and hence we require the pressure to fall continuously to zero at the edge of the contact disc, i.e. $p(a) = 0$. This gives

$$\Delta = ka^2 \tag{2.109}$$

and enables equation (2.108) to be written as

$$p(r) = -\frac{4k}{\pi A}\sqrt{a^2 - r^2} \tag{2.110}$$

The equation imposing normal equilibrium is given by

$$P = -\int_0^a p(r) 2\pi r dr \tag{2.111}$$

which gives

$$a^3 = \frac{3PA}{8k} \tag{2.112}$$

These results may be written in a variety of different ways. For example, if the pressure distribution is expressed as

$$p(r) = -p_0\sqrt{1 - (r/a)^2} \tag{2.113}$$

then the peak contact pressure, p_0, is

$$p_0 = \frac{3P}{2\pi a^2} = \frac{1}{\pi}\sqrt[3]{24P\left(\frac{k}{A}\right)^2} \tag{2.114}$$

The normal approach of distant points within the bodies is given by

$$\Delta = \frac{3PA}{8a} \tag{2.115}$$

or

$$\Delta^3 = \left(\frac{9A^2k}{64}\right)P^2 \tag{2.116}$$

and the normal stiffness (or the compliance) of the contact may therefore be evaluated. The next stage in the solution is to determine the internal stress field. This may be done by substituting the expression for $p(r)$ (equation 2.113) into equation (2.103) and evaluating the integral. This gives

$$g(t) = h(t)/f = -p_0 t \tag{2.117}$$

The complete stress field for a general point has been found before (Hamilton and Goodman, 1966, Hamilton, 1983, Sackfield and Hills, 1983, Hills, Nowell and Sackfield, 1993). Here, we shall record only the surface values. First, the stresses induced by the normal pressure, within the contact disc are

$$\frac{\sigma_{xx}^n}{p_0} = 2\nu K_0 + (1-2v)(G_0/r^2 - 2x^2G_0/r^4 + x^2K_0/r^2) \tag{2.118}$$

$$\frac{\sigma_{yy}^n}{p_0} = 2\nu K_0 + (1-2v)(G_0/r^2 - 2y^2G_0/r^4 + y^2K_0/r^2) \tag{2.119}$$

$$\frac{\tau_{xy}^n}{p_0} = (1-2\nu)(xyK_0/r^2 - 2xyG_0/r^4 \tag{2.120}$$

where

$$G_0 = a^2(1-(r/a)^2)^{\frac{3}{2}}/3$$

$$K_0 = -(1-(r/a)^2)^{\frac{1}{2}}$$

whilst those exterior to the contact disc are

$$\frac{\sigma_{xx}^n}{p_0} = \frac{a^2}{3}(1-2\nu)(2x^2/r^4 - 1/r^2) \tag{2.121}$$

$$\frac{\sigma_{yy}^n}{p_0} = \frac{a^2}{3}(1-2\nu)(2y^2/r^4 - 1/r^2) \tag{2.122}$$

$$\frac{\tau_{xy}^n}{p_0} = \frac{2a^2}{3}(1-2\nu)xy/r^4 \tag{2.123}$$

Now consider the influence of shearing traction. The stress field (other than traction) within the contact is

$$\frac{\sigma^t_{xx}}{fp_0} = -\frac{(4+\nu)\pi x}{8a} \tag{2.124}$$

$$\frac{\sigma^t_{yy}}{fp_0} = -\frac{3\nu\pi x}{8a} \tag{2.125}$$

$$\frac{\tau^t_{xy}}{fp_0} = -\frac{(2-\nu)\pi y}{8a} \tag{2.126}$$

and exterior to the contact is

$$\frac{\sigma^t_{xx}}{fp_0} = -(x/r^4)\left[2(r^2+\nu y^2)F_0 + \nu(3-4x^2/r^2)H_0\right] \tag{2.127}$$

$$\frac{\sigma^t_{yy}}{fp_0} = -(\nu x/r^4)\left[2x^2F_0 + (1-4y^2/r^2)H_0\right] \tag{2.128}$$

$$\frac{\tau^t_{xy}}{fp_0} = -(y/r^4)\left[(r^2-2\nu x^2)F_0 + \nu(1-4x^2/r^2)H_0\right] \tag{2.129}$$

where

$$F_0 = -\frac{1}{2}a\sqrt{(r/a)^2-1} + \frac{1}{2}(r^2/a)\tan^{-1}\left(1/\sqrt{(r/a)^2-1}\right)$$

$$H_0 = \frac{1}{2}a^3((r/a)^2-1)^{\frac{3}{2}} - \frac{1}{4}(r^4/a)\tan^{-1}\left(1/\sqrt{(r/a)^2-1}\right) - \frac{1}{4}ar^2\sqrt{(r/a)^2-1}$$

The surface displacements induced in each body by the contact pressure itself are the same if they have the same elastic constants, and hence there is no relative slip induced. In Chapter 4 we discuss the problems of slip induced by a mismatch of elastic constants more fully. However, the surface displacements produced by the shearing tractions are clearly unequal. They may be found from equations (2.85) to (2.87) and were given explicitly by Johnson (1955). They are

$$\frac{u}{fp_0} = \frac{\pi}{32\mu a}\left\{4(2-\nu)a^2 + (4-\nu)x^2 + (4-3\nu)y^2\right\} \qquad r < a \tag{2.130}$$

$$\frac{v}{fp_0} = \frac{\pi 2\nu xy}{32\mu a} \qquad r < a \tag{2.131}$$

2.5.2 Indentation by a rigid flat-ended punch

Indentation by a circular flat-ended punch produces a complete contact. As with the corresponding plane problem, the solution given is valid only if the punch is rigid so that the contacting body accommodates all the imposed deformation. As the punch has a square end, the deformation generated is constant, and hence

$$h(r) = \Delta \tag{2.132}$$

From equation (2.82) the transformed displacement h^* is also Δ, and evaluation of integral (2.83) gives

$$p(r) = -\frac{2\Delta}{\pi A a}\frac{1}{\sqrt{1-(r/a)^2}} \tag{2.133}$$

The equation of normal equilibrium (2.111) then gives

$$\Delta = \frac{PA}{4a}$$

so that

$$p(r) = -\frac{P}{2\pi a^2}\frac{1}{\sqrt{1-(r/a)^2}} \tag{2.134}$$

From here we can readily evaluate the kernels for the state of stress (equation 2.103) which gives

$$g(t) = h(t)/f = -\frac{P}{2\pi a^2}\delta\left(1-\frac{t}{a}\right) \tag{2.135}$$

where $\delta(.)$ is the Dirac delta function. The displacement field interior to the punch produced by a shear traction distribution $fp(r)$, i.e. when sliding occurs, is given by

$$\begin{aligned} u &= \frac{-P(2-\nu)}{8\mu a} \\ v &= 0 \end{aligned} \tag{2.136}$$

Note that a **normal** surface displacement also occurs which will only vanish if Dundurs' constant is zero, which for contact with a rigid punch means that the contacting body must be incompressible ($\nu = \frac{1}{2}$).

2.5.3 Indentation by a cone

Conical indentation provides a well-defined incomplete contact which shows potential as a fretting fatigue test geometry. The singularity of pressure at the apex will in practice be relieved by plasticity and also occurs within a stick zone; there is therefore the possibility of using this geometry as a means of verifying predictions made from a standard Hertzian contact. With the usual proviso that the cone angle $\pi/2-\theta$, must be large, we see that $h(r)$ is given by

$$h(r) = \Delta - \theta r \tag{2.137}$$

and from equation (2.4.2)

$$h^*(r) = \Delta - \frac{1}{2}\theta\pi r \tag{2.138}$$

Equation (2.83) then gives

$$p(r) = -\frac{2}{\pi A}\left\{\frac{\Delta}{\sqrt{a^2-r^2}} + \frac{\theta\pi}{4}\left[2\cosh^{-1}\left(\frac{a}{r}\right) - \frac{2a}{\sqrt{a^2-r^2}}\right]\right\} \tag{2.139}$$

If the possibility of adhesion if rejected $p(a) = 0$ and hence

$$\Delta = \frac{\pi\theta a}{2} \tag{2.140}$$

i.e.

$$p(r) = -\frac{\theta}{A}\cosh^{-1}\left(\frac{a}{r}\right) \tag{2.141}$$

Imposing normal equilibrium (equation 2.111) gives

$$a^2 = \frac{PA}{2\pi\theta} \tag{2.142}$$

and hence

$$p(r) = -\frac{\theta}{A}\cosh^{-1}\left\{\sqrt{\frac{PA}{2\pi\theta r^2}}\right\} \tag{2.143}$$

$$\Delta = \frac{PA}{4a} = \sqrt{\frac{PA\pi\theta}{8}} \tag{2.144}$$

The corresponding kernel for the internal stress field is

$$g(t) = h(t)/f = \frac{\theta}{A} \tag{2.145}$$

2.5.4 Indentation by a punch producing constant pressure

Once again this configuration is of practical interest as a soft punch may wear to a profile exerting a constant pressure. This will simply be given by

$$p(r) = \frac{P}{\pi a^2} \tag{2.146}$$

and substitution into equation (2.81) gives a transformed displacement

$$h^*(r) = -\frac{PA}{\pi a}\sqrt{1-(r/a)^2} \tag{2.147}$$

Equation (2.82) now constitutes an Abel equation whose inverse is

$$h(r) = -\frac{2PA}{\pi^2 a}E(r/a) \tag{2.148}$$

Here, $E(.)$ is the complete elliptic integral of the first kind. The stiffness of the contact is found from the central deflection, i.e.

$$h(0) \equiv \Delta = \frac{PA}{\pi a} \tag{2.149}$$

Lastly, the kernels for the internal stress field are

$$g(t) = h(t)/f = -\frac{2P}{\pi^2 a^2}\frac{t}{\sqrt{1-t^2}} \tag{2.150}$$

Explicitly, the surface displacements induced by the shear component of traction are found from equations (2.85) and (2.86) i.e.

$$2\mu u = \frac{2Pf}{\pi a}(\nu T^0_{xx} + T^0_{zz})$$

$$2\mu v = \frac{2Pf}{\pi a}vT^0_{xy}$$

In turn, the derivatives required are given by equations 3, 14, 15 of Table 2.2. The integrals required are

$$I^0_9 = \frac{2}{\pi}E(r/a) \tag{2.151}$$

$$I^0_{10} = \frac{a^2}{3r^2}\left\{(2-(r/a)^2)E(r/a) - (1-(r/a)^2)K(r/a)\right\} \tag{2.152}$$

$$I^0_{11} = -\frac{2a^4}{3r^4}\left\{(2-(r/a)^2)E(r/a) - (1-(r/a)^2)K(r/a)\right\} \tag{2.153}$$

Table 2.2 - Potentials needed for axi-symmetric problems

1.
$$N_x = T_{xz}/f = -xI_1$$

2.
$$N_{xz} = T_{xzz}/f = xI_3$$

3.
$$N_z = T_{zz}/f = I_9$$

4.
$$N_y = -yI_1$$

5.
$$N_{zz} = T_{zzz}/f = I_8$$

6.
$$N_{yz} = T_{yzz}/f = yI_3$$

7.
$$N_{xxz} = T_{xxzz}/f = I_3 - x^2(I_4 + I_5)$$

8.
$$N_{yyz} = I_3 - y^2(I_4 + I_5)$$

9.
$$N_{xyz} = T_{xyzz}/f = -xy(I_4 + I_5)$$

10.
$$N_{xzz} = -xI_6$$

11.
$$N_{yzz} = -yI_6$$

12.
$$N_{zzz} = -I_7$$

13.
$$N_{xx} = T_{xxz}/f = -I_1 + x^2 I_2$$

14.
$$T_{xx}/f = -\frac{1}{2}(I_9 + I_{10}) - \frac{x^2}{2}I_{11}$$

15.

$$T_{xy}/f = -\frac{1}{2}xyI_{11}$$

16.

$$T_{xyz}/f = xyI_2$$

17.

$$T_{xxx}/f = -\frac{3}{2}xI_{11} + x^3 I_{12}$$

18.

$$T_{xxxz}/f = 3xI_2 - x^3(2I_{13} + I_{14})$$

19.

$$T_{xyyz}/f = xI_2 - xy^2(2I_{13} + I_{14})$$

20.

$$T_{xxy}/f = -\frac{1}{2}yI_{11} + x^2 yI_{12}$$

21.

$$T_{xxyz}/f = yI_2 - x^2y(2I_{13} + I_{14})$$

where

$$I_1 = -\int_0^a \frac{g(t)\sin(\beta)}{q}\,dt$$

$$I_2 = -\int_0^a \frac{g(t)\sin(\alpha/2 + 2\beta)}{q^2\rho}\,dt$$

$$I_3 = -\int_0^a \frac{g(t)\sin(\alpha/2 + \beta)}{q\rho}\,dt$$

$$I_4 = -\int_0^a \frac{g(t)\sin(3\alpha/2 + \beta)}{q\rho^3}\,dt$$

$$I_5 = -\int_0^a \frac{g(t)\sin(\alpha + 2\beta)}{q^2\rho^2}\,dt$$

$$I_6 = -\int_0^a \frac{g(t)\sin(3\alpha/2)}{\rho^3}\,dt$$

$$I_7 = \int_0^a \frac{g(t)t\cos(3\alpha/2)}{\rho^3}\,dt + +zI_6$$

$$I_8 = -\int_0^a \frac{g(t)\sin(\alpha/2)}{\rho}\,dt$$

$$I_9 = \int_0^a g(t)\beta.dt$$

$$I_{10} = \int_0^a \frac{g(t)(t\cos\beta - z\sin\beta)}{q}\,dt$$

$$I_{11} = -\int_0^a \frac{g(t)\sin(2\beta)}{q^2}\,dt$$

$$I_{12} = -\int_0^a \frac{g(t)\sin(\alpha/2 + 3\beta)}{q^3\rho}\,dt$$

$$I_{13} = -\int_0^a \frac{g(t)\sin(\alpha + 3\beta)}{q^3\rho^2}\,dt$$

$$I_{14} = -\int_0^a \frac{g(t)\sin(3\alpha/2 + 2\beta)}{q^2\rho^3}\,dt$$

and

$$a^4\rho^4 = (r^2 + z^2 - t^2)^2 + 4z^2t^2$$

$$a^2q^2 = \{z + \rho\cos(\alpha/2)\}^2 + \{t + \rho\sin(\alpha/2)\}^2$$

$$\tan(\alpha) = \frac{2zt}{r^2 + z^2 - t^2} \qquad 0 \le \alpha \le \pi$$

$$\tan(\beta) = \frac{t + \rho\sin(\alpha/2)}{z + \rho\cos(\alpha/2)} \qquad 0 \le \beta \le \pi$$

Chapter 3

Contacts under Partial Slip

3.1 Introduction

Two fundamental features are present in any fretting problem: first, the contact must be experiencing shear tractions and this normally means that a resultant tangential force is transmitted from one body to another. Secondly, there must be some degree of relative tangential displacement between the contacting surfaces so that slip takes place during cyclic variation of the applied load. In Chapter 2, two types of contact were analysed:

(i) Those where only normal force was applied and no shear tractions arose.

(ii) Those where the shear tractions were everywhere equal to the limiting frictional value (i.e. $q(x, y) = fp(x, y)$). In this type of contact it follows that the total shear force, Q, is also equal to the limiting frictional value ($Q = fP$) and therefore gross sliding takes place between the contacting bodies.

It will be shown in Chapter 4 that it is possible to generate shear tractions from a purely normal force when the contacting bodies are elastically dissimilar, but nevertheless most practical fretting problems encountered will involve the transmission of a net shear force. Thus contacts of type (i) serve only as a useful introduction to the analysis of fretting problems. When analysing practical problems it will be necessary to account for the presence of shear tractions. Contacts of type (ii) are often employed in experimental investigations of fretting because it is relatively straightforward to set up the testing apparatus. However most fretting problems encountered in practice (such as bolted or riveted connections or dovetail roots of turbine blades) do not involve gross sliding and it is therefore important to examine the class of contacts where a shear force less than the limiting frictional value exists.

It is helpful to commence discussion of this subject by introducing the commonly accepted laws of frictional contact. The set of relationships commonly termed Coulomb friction were formulated by Amontons (1699) and Coulomb (1785) based on

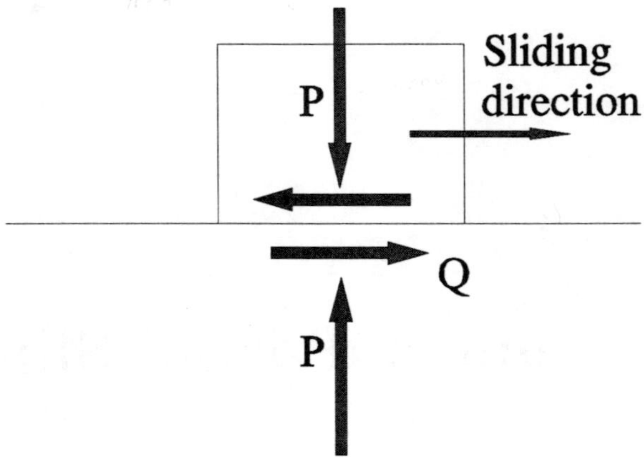

Figure 3.1: Gross sliding between rigid bodies.

experimental investigations of contact between rigid bodies. For the situation shown in fig. 3.1, where there is gross sliding between two *rigid* bodies

(i) The frictional force, Q, is proportional to the normal force, P, between the bodies.

$$Q = fP \tag{3.1}$$

where f is known as the coefficient of friction. Further, the frictional force acts in a direction which opposes the relative motion between the bodies.

(ii) The frictional force, Q, is independent of the apparent area of contact. Thus, a block with different sized faces will experience the same frictional force whichever face is in contact with another surface if the same normal force is applied and the surfaces have the same texture.

(iii) The frictional force occurring during gross sliding is independent of the velocity of sliding.

These three experimental observations form the basis of a simple model of friction. It must be emphasised that not all contacting bodies conform to this model. There would be no reason for specifying wide tyres on racing cars if observation (ii) applied to rubber. It is also important to remember that this frictional model applies to *sliding* contact of *rigid* bodies. However, an extension of the model is usually made to enable the analysis of frictional elastic contacts. It is convenient to examine this extension by reference to a two dimensional contact of the type discussed in §2.2 where surface tractions $p(x)$ and $q(x)$ are functions of a single ordinate, x. The application of a similar model to fully three dimensional contacts requires only a few modifications to the formulation.

Let us consider interaction between surfaces at a particular point, x. At this point the surfaces will either be slipping relative to each other or stuck. For the case of slipping the tractions are related by the friction law, so that

$$|q(x)| = -fp(x) \tag{3.2}$$

(the minus sign is introduced because of the convention employed that compressive surface tractions are negative). Further, the direction of the shear tractions opposes relative motion of the surfaces, giving

$$\mathrm{sgn}(q(x)) = -\mathrm{sgn}\left(\frac{\partial g}{\partial t}\right) \tag{3.3}$$

where $g = u_1(x) - u_2(x)$ is the relative displacement of the contacting surfaces and t is time. At first sight it might seem strange to introduce a time-derivative into a model for a quasi-static problem. However it should be apparent that if the applied load is truly static there will be no change in surface displacements and the particles cannot be said to be slipping. For the case where there is no relative slip at the point under consideration the shear tractions must be less than or equal to the limiting frictional value

$$|q(x)| \leq -fp(x) \tag{3.4}$$

Equations (3.2) to (3.4) give the usual interpretation of Coulomb friction employed in the analysis of elastic contacts where the friction law is taken to apply at each individual point within the contact rather than to a rigid body as a whole.

It is now possible to use this model to analyse elastic contacts where a shear force is applied which is less than the limiting frictional value. In general such contacts will be divided into zones of slip, where equations (3.2) and (3.3) apply, and zones of stick, where the shear tractions are less than the limiting value (3.4). Such a contact is said to be in a state of *Partial Slip.* In the stick zones there can be no relative motion and pairs of particles must remain adhered. Any difference in tangential displacement present when the pair of particles entered the stick zone must remain locked in; they can now only move together. Hence the relative displacement of pairs of surface particles must remain constant, i.e.

$$u_1(x) - u_2(x) = g(x) = g_0(x) \tag{3.5}$$

where $g_0(x)$ is the value of $g(x)$ when the particles entered the stick zone. In differential form this can be expressed as

$$\frac{\partial u_1(x)}{dx} - \frac{\partial u_2(x)}{dx} = \frac{\partial g(x)}{dx} \equiv g_0'(x) \tag{3.6}$$

and this gives a further condition which must be satisfied along with (3.4). For elastic contacts, equation (2.21) can be used to evaluate the value of $\frac{\partial g}{\partial x}$ in terms of the surface tractions. In this chapter we will confine our attention to contacts between elastically similar materials. In this case Dundurs' parameter, β (equation (2.18)), will be zero and therefore equation (2.21) simplifies to

$$\frac{1}{A}\frac{\partial g}{\partial x} = \frac{1}{\pi}\int_{contact} \frac{q(\xi)}{x-\xi}d\xi \tag{3.7}$$

It will be noticed that this equation is now uncoupled in the sense that it contains only $q(x)$ and not $p(x)$. We shall now proceed to obtain solutions to equation (3.7) for cases of partial slip (i.e. $Q < fP$) for a range of different geometries and loading histories. The geometries examined might at first seem to be rather idealised and restrictive. There are two reasons for this: first the analysis of practical geometries is often extremely complicated and is best undertaken numerically (e.g. by the finite element method). It is difficult to have confidence in the solutions obtained unless simple geometries are analysed first in order to calibrate the numerical technique. Secondly, experimental investigations of fretting fatigue are best carried out under controlled and repeatable conditions. This usually involves the use of a contact configuration which is 'incomplete' in nature (see §2.1) and it is this type of contact which we will consider here.

3.2 Contact of cylinders under partial slip

We begin a discussion of partial slip configurations by addressing the essentially two-dimensional case of the contact of two elastically-similar cylinders, as analysed for normal loading alone in §2.2. When the cylinders (or a cylinder and a half-plane) are pressed together under the action of a normal force, P, a Hertzian contact of semi-width a occurs with an elliptical distribution of normal pressure given by equation (2.37)

$$p(x) = -p_0\sqrt{1-(x/a)^2} \tag{3.8}$$

Let us now consider the *subsequent* application of a monotonically increasing tangential force, Q, less than the limiting frictional value ($Q < fP$), as shown in fig. 3.2. It is anticipated that the contact will be divided into zones of slip and zones of stick, but one of the difficulties of analysing frictional contacts is that the distribution of stick and stick zones is unknown *a priori.* Although it is possible to develop methods of analysis which automatically determine the stick/slip regime (e.g. Kalker, 1990), for simple configurations it is usually preferable to guess the extent of the stick and slip zones and verify afterwards that the appropriate conditions (equations (3.2) to (3.4) and (3.6)) are satisfied.

It is often instructive to examine first the case where the coefficient of friction is sufficiently high to prevent all relative slip. Locations where the implied ratio $q(x)/p(x)$ is high will then suggest where slip is likely to occur if the coefficient of

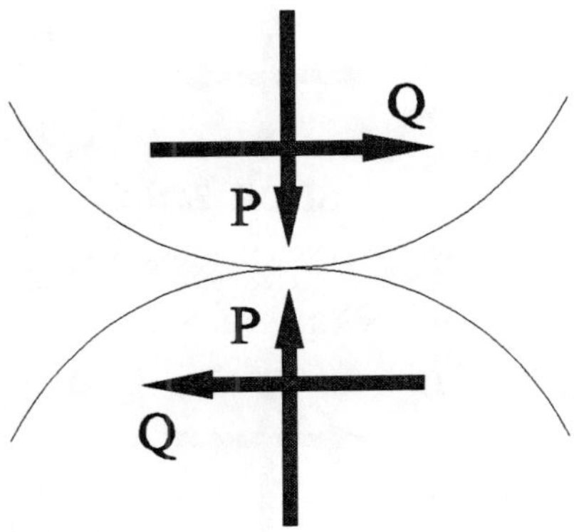

Figure 3.2: Contact of cylinders under the action of a normal force together with a subsequently applied shear force.

friction is reduced. We shall therefore commence the analysis of the contacting cylinders by making this assumption of no slip. Before the application of the tangential force no shear tractions will have arisen and there will have been no relative motion between the surfaces. Thus $g_0'(x) = 0$ and, since the stick zone encompasses the entire contact $|x| \leq a$, the shear tractions may be determined from (3.6) and (3.7)

$$\frac{1}{\pi}\int_{-a}^{a}\frac{q(\xi)}{x-\xi}d\xi = 0 \qquad |x| \leq a \tag{3.9}$$

The solution (or *inversion*) of this equation may be obtained in a similar manner to that for equation (2.22). In this case, however, the right hand side of equation (3.9) is zero and hence the problem is identical in form to that arising in normal indentation by a flat punch (equations (2.64) and (2.22)). The solution to this type of equation has already been found as (2.66)

$$q(x) = \frac{C}{\sqrt{a^2 - x^2}} \tag{3.10}$$

where C is a constant, the value of which may be found by considering overall tangential equilibrium

$$\int_{-a}^{a} q(x)dx = Q \tag{3.11}$$

which gives

$$C = \frac{Q}{\pi} \tag{3.12}$$

It will be immediately apparent that $q(x)$ approaches infinity as $x \to \pm a$ whereas the normal pressure (3.8) falls to zero. Thus the ratio $q(x)/p(x) \to \infty$ as $x \to \pm a$ so that an infinite coefficient of friction is required to prevent slip. It is therefore

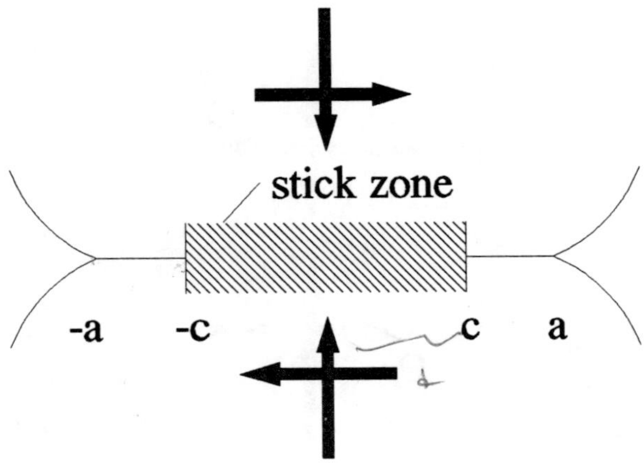

Figure 3.3: Stick and slip regions for the contact of cylinders in partial slip.

inevitable that some slip will occur if a cylindrical Hertzian contact is loaded in the manner described even if the shear force is very small. As noted by Johnson (1985), this conclusion is hardly surprising since the assumption of no slip requires the two bodies to behave as one. The edges of the contact appear, under this assumption, to be the tips of two long sharp cracks loaded in shear and we would expect stress singularities to arise at these points.

The existence of high shear tractions at either edge of the contact in the no-slip solution suggests that in practice slip may take place here. We will therefore begin our analysis of the partial slip case by postulating that slip takes place in two symmetrical regions $a > |x| \geq c$ which surround a central stick zone $|x| < c$ (fig. 3.3). Note that the symmetry of the loading, geometry and material constants about the x and y axes also suggests that the stick zone will be symmetrical about the centre of the contact. It is convenient to model the shear tractions as a perturbation on the fully-sliding solution so that

$$q(x) = f p_0 \sqrt{1-(x/a)^2} + q'(x) \tag{3.13}$$

where the perturbation $q'(x) = 0$ in the slip zones ($a > |x| > c$). This automatically ensures that one of the conditions for slip (3.2) is satisfied in the slip regions. It remains to determine $q'(x)$ in the stick zone and the extent of this zone, c. Once again, there is no relative displacement in the stick zone so that $g_0'(x) = 0$ and hence

$$\frac{1}{\pi}\int_{-a}^{a} \frac{q(\xi)}{x-\xi} d\xi = 0 \qquad |x| \leq c \tag{3.14}$$

Substitution of (3.13) into (3.14), followed by normalisation of the integral over $|x| < c$ and integration of the sliding term yields

$$\frac{1}{\pi}\int_{-1}^{1} \frac{q'(v)dv}{v-t} = f\ p_0 \left(\frac{c}{a}\right) t \qquad -1 < t < 1 \tag{3.15}$$

where $v = \xi/c$ and $t = x/c$. Once again this equation may be inverted using the general recipe of (2.23). This time, however, the form of the equation is similar to that arising in the normal contact of cylinders (§2.3.1) so that a bounded fundamental function (equation 2.32) is appropriate and the solution becomes

$$q'(t) = -fp_0(c/a)\sqrt{1-t^2} \qquad |t| < 1 \tag{3.16}$$

It remains to determine the stick zone size, c. This may be effected by enforcing tangential equilibrium so that

$$\begin{aligned} Q &= \int_{-a}^{a} q(\xi)d\xi \\ &= fp_0 a \int_{-1}^{1} \sqrt{1-s^2}ds - fp_0\frac{c^2}{a}\int_{-1}^{1}\sqrt{1-s^2}ds \end{aligned} \tag{3.17}$$

which gives

$$Q = \frac{fp_0\pi}{2a}\left(a^2 - c^2\right) \tag{3.18}$$

This equation may be cast in a more convenient form by combining it with (2.38) to give

$$c/a = \sqrt{1 - |Q/fP|} \tag{3.19}$$

We must now check that the stick and slip conditions are satisfied in the appropriate regions. The solution has been obtained by enforcing (3.6) in the stick zone and it is clear that (3.4) is satisfied in the stick region since the corrective term, $q'(x)$, is opposite in sign to the fully sliding term. Equation (3.2) has, as was remarked above, been automatically satisfied in the slip zones. The requirement for consistency of slip direction and shear tractions (3.3) needs to be verified for a monotonically increasing load, Q. This can be achieved by examining the relative displacements using the results of Poritsky (1950) or the technique outlined later in this section. Figure 3.4 depicts the distribution of shear tractions obtained for different values of Q/fP. The central stick zone where the shear tractions are reduced in magnitude is immediately apparent.

The classical solution just determined is often known as the Mindlin solution since the corresponding (and very similar) axi-symmetric case was analysed by Mindlin (1949), using an alternative method of analysis. So far we have merely determined the surface traction distribution. The next step is to calculate the corresponding stress and displacement fields. It will be apparent that the surface tractions can be thought of as a superposition of three elliptical distributions:

(i) A distribution of normal pressure of peak magnitude $-p_0$ acting between $x = -a$ and $x = a$.

(ii) A distribution of shear traction of peak magnitude fp_0 acting between $x = \pm a$.

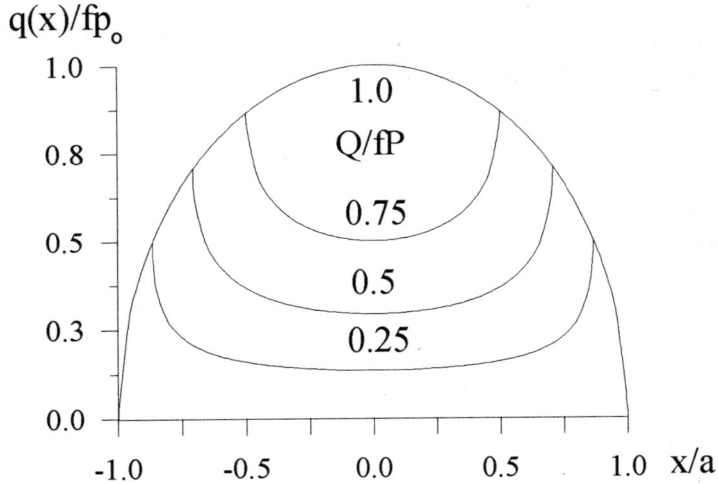

Figure 3.4: Shear traction distributions for a range of values of tangential force, Q.

(iii) A second shear traction distribution of peak magnitude $-fp_0c/a$ between $x = \pm c$.

Since the analysis is linear-elastic it is possible to determine the stress and displacement fields by appropriate superposition of the results for (i), (ii), and (iii). For example, to calculate the σ_{xx} component of stress at a point (x, y) (in un-normalised coordinates) for a contact transmitting normal and shear tractions and for which the stick zone semi-width is c, it is merely necessary to superimpose three results to give

$$\sigma_{xx}(x,y) = p_0\left(\frac{\sigma^n_{xx}\left(\frac{x}{a},\frac{y}{a}\right)}{p_0}\right) + fp_0\left(\frac{\sigma^t_{xx}\left(\frac{x}{a},\frac{y}{a}\right)}{fp_0}\right) - fp_0\frac{c}{a}\left(\frac{\sigma^t_{xx}\left(\frac{x}{c},\frac{y}{c}\right)}{fp_0}\right) \tag{3.20}$$

where the functions in brackets are given by (2.44). To obtain the equivalent result for a shear force applied in the opposite direction it is merely necessary to change the sign of the second and third terms. Similar superposition may be carried out to determine the displacements. Surface values are particularly important in fretting fatigue since they allow the calculation of relative slip and these may be obtained by superposition using equations (2.57) - (2.63).

It is sometimes useful to calculate the normal and tangential compliances of contacting cylinders (i.e. the ratios δ_n/P and δ_t/Q, where δ_n and δ_t represent the displacement between a point at the centre of the contact and a distant point in the body in the normal and tangential directions respectively). Unfortunately the expressions for displacements within *plane* bodies (see equations (2.57) to (2.63), Poritsky (1950), or Hills, Nowell and Sackfield (1993)) do not reach a limiting value remote from the contact and thus the compliances defined above are infinite. This is clearly not the case in a practical situation. However all practical situations are actually three dimensional and the two-dimensional model analysed here is merely an approximation. It is possible to evaluate compliances for three-dimensional configurations and hence the difficulty experienced here may be considered an artefact of the two-dimensional model employed. We shall analyse a three dimensional contact next, before discussing the general treatment of the case where an applied tangential load is subsequently reduced (§3.4). This last problem is of interest in both two and three dimensional configurations.

3.3 Contact of spheres under partial slip

One of the difficulties involved in employing cylindrical pads in fretting fatigue experiments is ensuring that the axis of each pad is exactly parallel to the surface of the specimen in order that the surface tractions do not vary along the length of the contact patch[1]. This difficulty can be minimised by the use of pressure sensitive film and careful setting-up (Nowell, 1988) but it is sometimes attractive to use spherical fretting pads, where no such alignment difficulties exist (Kuno et al., 1989). It is therefore necessary to examine the effect of a tangential force on the tractions present when a sphere is in contact with a flat surface (or, more generally, two spheres are in contact). If the materials are elastically similar, Dundurs' parameter, β, is zero and the normal pressure remains Hertzian as derived in §2.5.1. The contact patch is a circle of radius a and the pressure is given by

$$p(r) = -p_0\sqrt{1-(r/a)^2} \tag{3.21}$$

The configuration is, of course, the three-dimensional analogue of that studied in §3.2. However, the axi-symmetric nature of the problem disappears as soon as a tangential force, Q, is applied (fig. 3.5). Once again, useful insight into the likely distribution of stick and slip zones can be gained by examining first the problem where no slip is permitted. It is convenient to take the z-axis as normal to the contact patch and the x and y axes in the plane of the contact (fig. 3.5). Consider a monotonically increasing force, Q, applied parallel to the x-axis. For no slip to take place anywhere within the contact area there must be no relative displacement and the x-direction displacement of all contacting points must be constant i.e.

$$u_1 = u_2 = \text{ constant} \qquad\qquad r \leq a \tag{3.22}$$

The distribution of *normal* traction required to produce uniform *normal* displacement over a circular area was derived when considering the problem of indentation of a half-space by a rigid circular punch (§2.5.2, equation 2.134) and it is clear that the equations describing the current problem are of a similar form. Therefore, by analogy,

[1]See Chapter7 for further discussion of alignment problems.

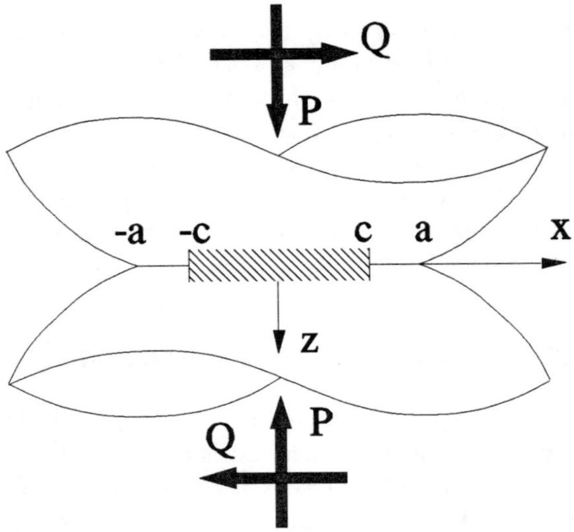

Figure 3.5: A circular Hertzian contact subjected to a shear force, Q.

it is possible to deduce that the required distribution of shear traction to cause a constant x-direction displacement over the contact circle takes the form (Hills, Nowell, and Sackfield, 1993)

$$q_x(r) = \frac{q_0}{\sqrt{1-(r/a)^2}} \tag{3.23}$$

The magnitude of this distribution is radially symmetric but the traction is directed in the x-direction only, as indicated by the subscript. The constant q_0 may be found as usual from considerations of tangential equilibrium so that

$$q_0 = \frac{Q}{2\pi a^2} \tag{3.24}$$

Comparing the form of equations 3.21 and 3.23 we see that, as with the problem of tangential loading of cylinders, an infinite coefficient of friction is required at the edge of the contact in order to prevent the onset of slip. It is also apparent that, if friction is finite, slip is most likely in an annular region around the edge of the contact disc.

We now move on to consider the partial slip problem. As remarked above, the axi-symmetric nature of the normal contact problem is removed once the tangential force is applied. This lack of obvious symmetry means that there is no simple way to formulate the problem directly. Instead it is convenient to pursue a superposition argument first advanced by Cattaneo (1938) and independently by Mindlin (1949). Consider an elliptical distribution of shear traction, acting in the x-direction, of sufficient magnitude to cause sliding.

$$q_x(r) = fp_0\sqrt{1-(r/a)^2} \qquad r < a \tag{3.25}$$

The x and y components of surface displacement in each body under this traction distribution may be found from equations (2.130) and (2.131) as

$$u = \frac{fp_0\pi a}{32\mu}\left[4(2-\nu) + (4-\nu)\left(\frac{x}{a}\right)^2 + (4-3\nu)\left(\frac{y}{a}\right)^2\right] \qquad \left(x^2+y^2\right) \le a^2 \quad (3.26)$$

and

$$v = \frac{fp_0\pi a\nu}{16\mu}\left(\frac{x}{a}\right)\left(\frac{y}{a}\right) \qquad \left(x^2+y^2\right) \le a^2 \quad (3.27)$$

There is clearly no stick region since the displacement components of adjacent particles differ. However, guided by the solution obtained for cylinders, we now consider a second, geometrically similar, shear traction distribution of opposite sign and reduced magnitude applied over the central region $r < c$, i.e.

$$q'_x = -fp_0\left(\frac{c}{a}\right)\sqrt{1-(r/c)^2} \qquad r < c \quad (3.28)$$

The net displacement field in the region $r < c$ due to both q and q' can be determined by suitably scaled superposition of (3.26) and (3.27) to give

$$u = \frac{fp_0\pi a}{8\mu}(2-\nu)\left[1-\left(\frac{c}{a}\right)^2\right] \qquad r \le c \quad (3.29)$$

$$v = 0 \qquad r \le c \quad (3.30)$$

Thus, both x and y displacements within the central region $r \le c$ are constant and equation (3.5) is satisfied. Once again the shear tractions here are clearly less than the limiting frictional value so that equation (3.4) is also satisfied and the region constitutes a stick zone. It remains to check the conditions in the slip annulus $a \ge r > c$. Here q' is zero and hence the x-direction shear traction (3.25) is limiting as required by (3.2). The requirement for the shear tractions to oppose relative displacement must also be tested. For three dimensional problems shear tractions and relative slip can be in any direction in the x-y plane. The restriction in the slip zones is more severe than equation (3.3) in that shear tractions and relative slip increment must be exactly opposite in direction. Thus, in the current configuration, where the shear tractions are all directed in the x-direction, it is necessary to check that the incremental slip has no y-component and that its x-component has the correct sign. To do this the surface displacements exterior to an elliptical distribution of shear traction applied over a circular region are required. These have been determined by Johnson (1955) as

$$\begin{aligned}\frac{8\mu}{afp_0}u &= (2-\nu)\left\{\left[2-\left(\frac{r}{a}\right)^2\right]\sin^{-1}\left(\frac{a}{r}\right) + \left(\frac{r}{a}\right)\sqrt{1-\left(\frac{a}{r}\right)^2}\right\} \\ &+\frac{1}{2}\nu\left\{\left(\frac{r}{a}\right)^2\sin^{-1}\left(\frac{a}{r}\right) + \left[2-\left(\frac{r}{a}\right)^2\right]\sqrt{1-\left(\frac{a}{r}\right)^2}\left(\frac{a}{r}\right)\right\}\left[\left(\frac{x}{a}\right)^2-\left(\frac{y}{a}\right)^2\right]\end{aligned} \quad (3.31)$$

$$\frac{8\mu}{afp_0}v = \nu\left\{\left(\frac{r}{a}\right)^2 \sin^{-1}\left(\frac{a}{r}\right) + \left[2 - \left(\frac{r}{a}\right)^2\right]\sqrt{1-\left(\frac{a}{r}\right)^2}\left(\frac{a}{r}\right)\right\}\frac{xy}{a^2} \tag{3.32}$$

For $r \geq a$. It will be apparent that, unless Poisson's ratio is zero, there is a y component of displacement in the slip annulus ($a \leq r < c$) arising from a superposition of (3.27) and (3.32). There will therefore be a y component of relative slip between two contacting spheres since the opposing directions of shear tractions on each sphere cause y components of displacement in opposite directions. As remarked above the shear tractions are all directed in the x-direction. Thus, the requirement for the shear tractions to oppose relative slip is not precisely satisfied. The errors involved are not large, but neither are they second order and they clearly increase in proportion to Poisson's ratio. A solution to the problem where the tractions are correctly aligned has been obtained using a numerical method by Munisamy, Hills, and Nowell (1992). The largest discrepancies between the two solutions occur on the lines $y = \pm x$ and the y component of slip can be up to 10% of the corresponding x component for materials with a Poisson's ratio of 0.5.

As with the plane problem, the size of the stick zone for spheres in partial slip must be determined by overall equilibrium so that

$$Q = \int_0^a 2\pi r q_x(r)dr = 2\pi f p_0\left[\int_0^a r\sqrt{1-(r/a)^2}\,dr - \frac{c}{a}\int_0^c r\sqrt{1-(r/c)^2}\,dr\right] \tag{3.33}$$

which yields, after combining with (2.114),

$$\frac{Q}{fP} = 1 - \left(\frac{c}{a}\right)^3 \tag{3.34}$$

or, more conveniently

$$\left(\frac{c}{a}\right) = \sqrt[3]{1 - \frac{Q}{fP}} \tag{3.35}$$

As with the two-dimensional case, the subsurface stress field may be obtained by superposition of the expressions for a fully-sliding contact since all the traction distributions are of the form $C\sqrt{1-(r/a)^2}$. The relevant equations for stresses at the surface are presented in Chapter 2 ((2.118)-(2.129). Subsurface values have been calculated by Hamilton and Goodman (1966) and are also presented by Hills, Nowell, and Sackfield (1993). Equations for the surface displacements have been given above (3.26)-(3.27) and (3.3)-(3.32). In contrast to the two-dimensional case of the contact of cylinders, it will be seen that it *is* possible to evaluate the tangential compliance of the contact. Equation (3.29) gives the displacement of points in the stick zone with respect to a distant point in the body. This expression is thus equal to the tangential compliance of half of the contact so that, after appropriate substitution and manipulation the following equation is obtained

$$\frac{\Delta}{Q} = \frac{3}{16a}\left\{\frac{2-\nu_1}{\mu_1} - \frac{2-\nu_2}{\mu_2}\right\}\left\{1 - [1 - Q/fP]^{\frac{2}{3}}\right\} \tag{3.36}$$

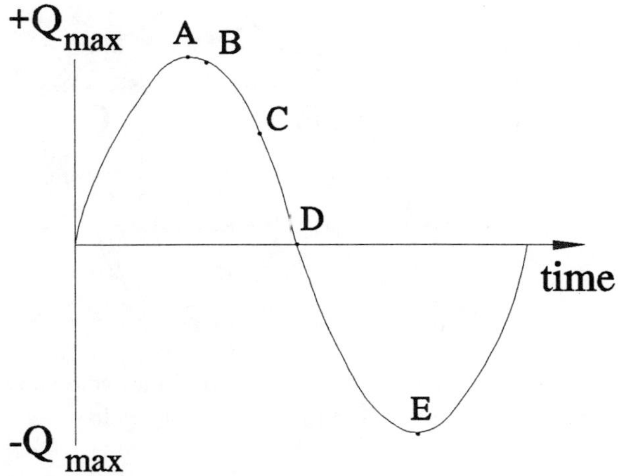

Figure 3.6: Tangential load history during fretting fatigue.

It should be noted that this expression is exact only when the elastic constants of the two contacting bodies are identical. This is because for dissimilar materials coupling affects the shear traction distribution (see Chapter 4).

3.4 Load variation and history dependence

The configurations analysed so far in this chapter have all been for a monotonically increasing tangential force. In fretting fatigue the cyclic variation of a tangential force is a prerequisite. If the tangential force is cycled between constant limits ($\pm Q_{\max}$, fig. 3.6), it transpires that the expressions developed so far are applicable when the applied force is at either of its extreme values. To show that this is the case or to examine the stresses or displacements at some other point in the fretting cycle it is necessary to examine more closely what happens on the reversal of load.

Consider, as an example, the case of contact of cylinders examined in §3.2. Initially the tangential load, Q, increases monotonically from 0 to $Q_{\max}$ (point A in fig. 3.6). The shear traction distribution occurring at this point will be precisely as described by equations (3.13) and (3.16) where the stick zone size, c, is given by

$$\frac{c}{a} = \sqrt{1 - \frac{Q_{\max}}{fP}} \tag{3.37}$$

Consider now the situation at point B (fig. 3.6) where the load has been reduced *infinitesimally* from its maximum value. The shear tractions reduce *infinitesimally* and there is thus a corresponding decrease in tangential displacement. Hence the rate of change of relative displacement $\frac{\partial g}{\partial t}$ is opposite in sign to that occurring during the

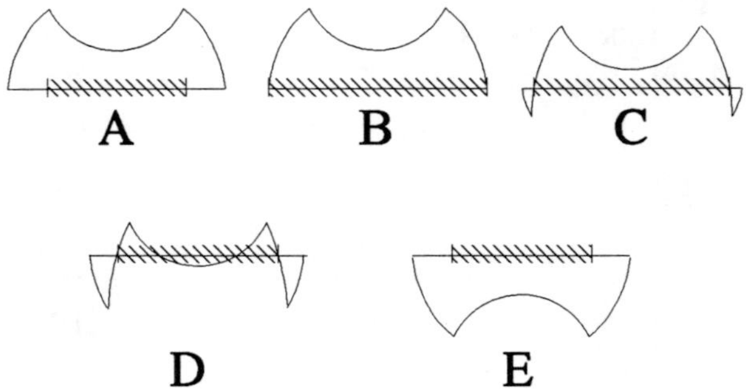

Figure 3.7: Shear tractions occurring during the loading cycle shown in Figure 3.6. The labels correspond to different points in the loading cycle.

period of increasing load. This means that the requirement for the shear tractions in the slip zones to oppose relative motion (3.3) is violated and instantaneous stick must occur over the entire contact. A further reduction in Q, to point C in fig. 3.6, must lead to *reverse* slip at the edges of the contact (fig. 3.7) since, by the arguments presented in §3.2, the normal pressure will be insufficient to prevent relative slip here. In these new slip zones ($c' < |x| \leq a$, say) the shear traction will have changed from $fp_0\sqrt{1-(x/a)^2}$ to $-fp_0\sqrt{1-(x/a)^2}$ and this on its own this will give rise to relative displacement elsewhere within the contact. An additional corrective traction distribution must therefore be applied over the new stick zone ($|x| \leq c'$) and, by analogy with the results of §3.2, it is possible to infer that the correction $q''(x)$ is given by

$$q''(x) = +2fp_0\frac{c'}{a}\sqrt{1-(x/c')^2} \tag{3.38}$$

The factor of two occurs because the corrective term must cancel the relative displacements occurring when the slip zone tractions change by $2fp_0\sqrt{1-(x/a)^2}$ instead of $fp_0\sqrt{1-(x/a)^2}$ as occurred during monotonically increasing loading. The total shear traction is therefore the sum of the four terms given below

$$\begin{array}{ccc}
q(x) & \text{zone of application} & \text{contribution to shear force} \\
fp_0\sqrt{1-(x/a)^2} & |x| \leq a & \frac{1}{2}fp_0\pi a \\
-fp_0\left(\frac{c}{a}\right)\sqrt{1-(x/c)^2} & |x| < c & -\frac{1}{2}fp_0\pi c^2/a \\
-2fp_0\sqrt{1-(x/a)^2} & |x| \leq a & -fp_0\pi a \\
2fp_0\left(\frac{c'}{a}\right)\sqrt{1-(x/c')^2} & |x| < c' & fp_0\pi(c')^2/a
\end{array} \tag{3.39}$$

This gives net shear tractions as follows

$$
\begin{aligned}
q(x)/fp_0 &= -\sqrt{1-(x/a)^2} && c' < |x| \le a \\
&= -\sqrt{1-(x/a)^2} + 2(c'/a)\sqrt{1-(x/c')^2} && c \le |x| \le c' \\
&= -\sqrt{1-(x/a)^2} + 2(c'/a)\sqrt{1-(x/c')^2} \\
&\quad -(c/a)\sqrt{1-(x/c)^2} && |x| \le c
\end{aligned}
\tag{3.40}
$$

As with the case of monotonic loading, the position of the stick zone boundary must be obtained from overall equilibrium so that

$$Q = \frac{fp_0\pi}{2a}\left(-a^2 + 2(c')^2 - c^2\right) \tag{3.41}$$

which yields

$$\left(\frac{c'}{a}\right)^2 = 1 - \frac{1}{2fP}(Q_{\max} - Q) \tag{3.42}$$

Thus, when the shear force has been completely removed (point D, fig. 3.6) it is found that

$$\left(\frac{c'}{a}\right)^2 = 1 - \frac{Q_{\max}}{2fP} = \frac{1}{2}\left(1 + (c/a)^2\right) \tag{3.43}$$

It is clear that non-zero but self-equilibrating shear tractions persist, even though the load has been removed. This illustrates an important aspect of the analysis of frictional contacts, namely that the traction distribution (and hence the resulting stresses and displacements in the contacting bodies) depends on the history of loading. Thus, in contrast to most other applications of elasticity, we cannot necessarily exploit superposition in determining the stress state. Proceeding with the example of the contact of cylinders it is found that when the load is completely reversed and $Q = -Q_{\max}$ (point E, fig. 3.6) the shear traction distribution is also completely reversed. Figure 3.8 depicts the history of shear tractions for a sample case. If the evaluation of stress and displacement fields at the extreme points in the loading cycle is all that is required (and this is often the case, for example in calculations of the stress intensity factor range ΔK (see Chapter 6) under fully reversing shear force) then the expressions given in §3.2 for monotonic loading are applicable. At other points during the loading cycle it is necessary to determine the shear traction distribution using equations (3.38) to (3.42) before establishing the stresses and displacements using superposition of elliptical distributions of normal and shear traction.

One parameter which is of interest in the analysis of fretting fatigue is the frictional energy dissipated in relative slip at the interface during cyclic loading. On a macroscopic scale the total energy expended can be obtained from a plot relating the applied shear force, Q, and the displacement between distant points, Δ. It is apparent

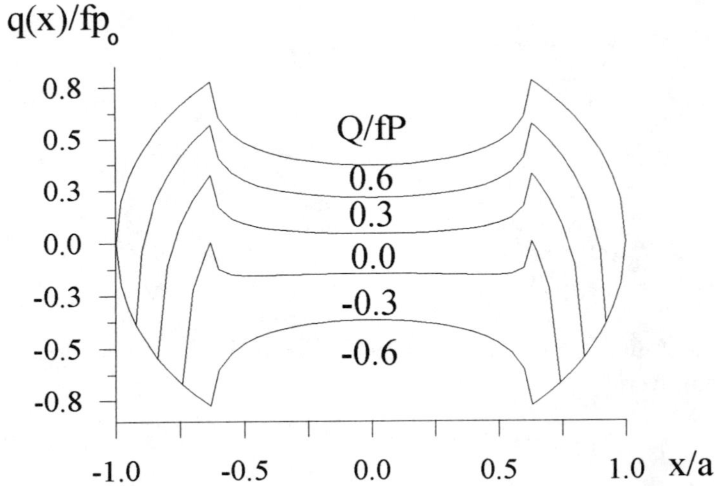

Figure 3.8: Shear traction distributions during cycling of tangential load, $Q_{\max}/fP = 0.6$.

from the example of the contact of spheres (equation 3.36) that, during monotonic loading, Δ is a non-linear function of Q. The same non-linearity occurs during cyclic loading and a hysteresis loop of the type shown in figure 3.9 is obtained. The total energy expended per cycle is given by the area enclosed within the loop. Perhaps of even greater interest, however, is the distribution of this energy expenditure within the contact on a microscopic scale. If the shear traction at a particular point, x, is plotted against the relative slip, $g(x)$, occurring between the surfaces at the same point, a rectangular hysteresis loop of the type shown in figure 3.10 is obtained. The energy dissipated per unit contact area is given by the area within this curve which is proportional to the product of the maximum relative displacement and maximum shear traction.[2] The calculation of the maximum value of $g(x)$ may be achieved by appropriate superposition of the equations for displacement (2.57) - (2.63). An alternative method is to integrate equation (3.7) from the appropriate edge of the smallest stick zone ($|x| < c$) since $g(x) = 0$ here, where particles have never slipped. Figure 3.11 shows a typical distribution of relative displacement occurring when Q is at a maximum value (point A in fig. 3.6) . Clearly if the frictional energy density calculated in this manner is integrated over the entire contact the total energy dissipated in one complete cycle, W, will be obtained and this will equal that calculated by the macroscopic approach, i.e.

$$W = \int_{\text{cycle}} Q\, d\Delta = 4 \int_{\text{contact}} q_{\max} g_{\max}\, dx \tag{3.44}$$

[2]This parameter is sometimes referred to as $\tau.\delta$, see Chapter 9.

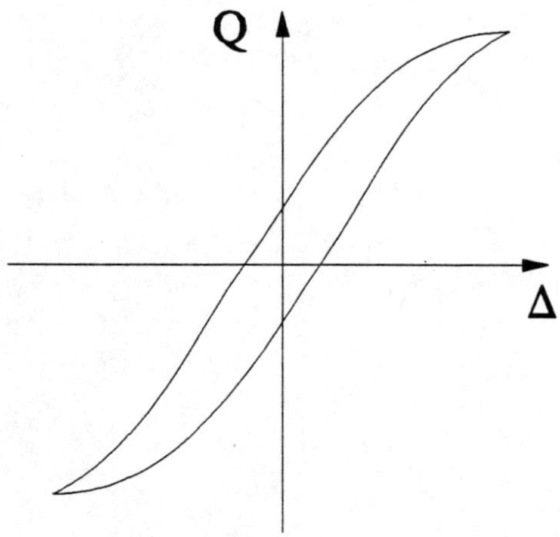

Figure 3.9: Hysteresis loop between tangential force, Q, and displacement of distant points, Δ, occurring during cyclic loading.

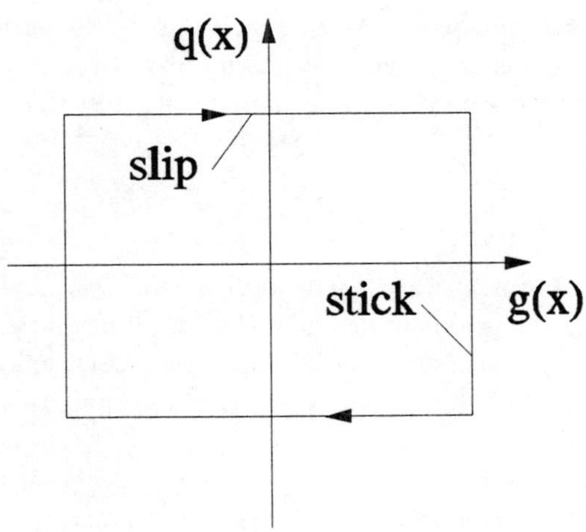

Figure 3.10: Hysteresis loop between shear traction, $q(x)$, and relative displacement, $g(x)$, occurring in the slip zone during cyclic tangential loading.

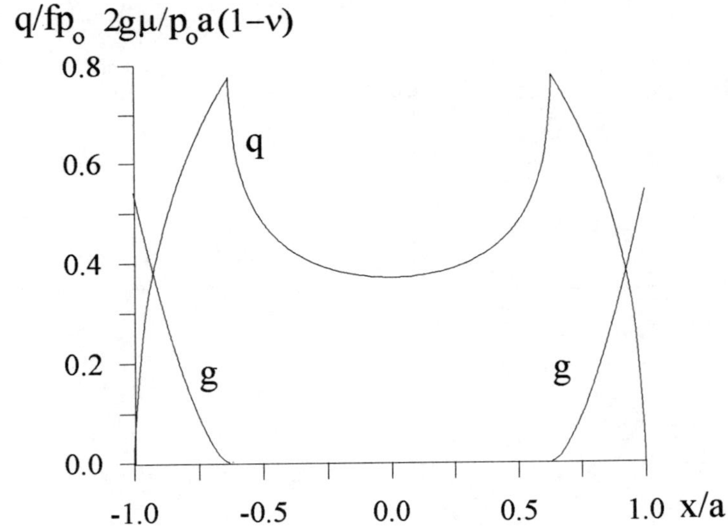

Figure 3.11: Shear tractions and relative displacement, $g(x)$, occuring at $Q = Q_{max}$ and $Q_{max}/fP = 0.6$.

where dx is replaced in the second integral by dA for three-dimensional contacts.

The traction distributions occurring during cyclic loading of other contact configurations can be determined in a similar manner to that described above. The case of cyclic tangential loading of contacting spheres was first analysed by Mindlin and Deresiewicz (1953). For this case reverse slip takes place over an annulus $r > c'$ during reversal of the load, where c' may be determined from overall equilibrium to give

$$\frac{c'}{a} = \sqrt[3]{1 - \frac{1}{2fP}\left(Q_{\max} - Q\right)} \tag{3.45}$$

and $Q_{\max}$ is the maximum forward load applied. In this case the distribution of frictional energy expended is axisymmetric if the Mindlin solution for shear tractions is used. Munisamy, Hills, and Nowell (1992) show that local maxima occur on $x = 0$ if the solution is modified to ensure that shear tractions precisely oppose relative slip in the slip region (see §3.3).

A further interesting example of the history dependence of frictional contacts is illustrated by the configuration shown in fig. 3.12. Here a pair of contacting cylinders is loaded by a monotonically-increasing inclined force, F, so that the normal and tangential components, P and Q, increase simultaneously and in proportion. It can be shown that, in contrast to the case examined in §3.2, where Q is applied after P, no slip takes place unless the contact enters the regime of gross sliding. To investigate this case it is convenient to begin with the assumption that no slip takes place and to check the resulting solution for consistency. It will be recognised that the contact

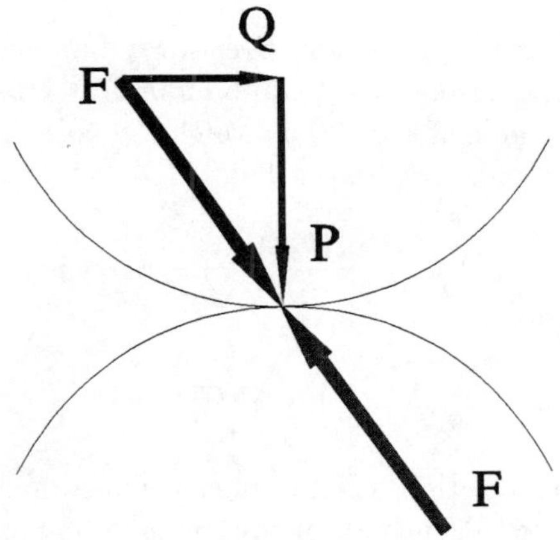

Figure 3.12: Contact of cylinders under a monotonically increasing inclined load.

area will increase during loading and the contact semi-width, a, can be related to the normal component of the load by (2.36)

$$a^2 = \frac{2AP}{\pi k} \tag{3.46}$$

Suppose an increment in load dF is made. The normal and tangential components of this increment are clearly

$$\begin{aligned} dP &= dF \cos\alpha \\ dQ &= dF \sin\alpha \end{aligned} \tag{3.47}$$

so that

$$dQ = dP \tan\alpha \tag{3.48}$$

The increase in contact semi-width, da, is related to the normal load increment and can be obtained by differentiating (3.46) so that

$$2a\, da = \frac{2A}{\pi k}\, dP \tag{3.49}$$

If no slip occurs in the contact as a result of the load increment, the incremental increase in shear traction must be that which causes no relative displacement. The required distribution has already been found in examining tangential loading of cylinders without slip (§3.2, equations 3.10 and 3.12) so that

$$dq(x) = \frac{dQ}{\pi\sqrt{a^2 - x^2}} \tag{3.50}$$

Equations (3.48) and (3.49) may now be substituted into (3.50) to give

$$dq(x) = \frac{ka \tan\alpha\, da}{A\sqrt{a^2 - x^2}} \tag{3.51}$$

The shear traction occurring at a point x, after several increments of load, when the contact has reached a given size, $a = a_1$, can be found by integrating the increments given by (3.51) from the value of a $(= x)$ at which the point entered the contact and hence started to 'acquire' shear tractions. Thus

$$q(x) = \frac{k \tan\alpha}{A} \int_x^{a_1} \frac{a}{\sqrt{a^2 - x^2}}\, da \tag{3.52}$$

so that

$$q(x) = \frac{k \tan\alpha}{A} \sqrt{a_1^2 - x^2} \tag{3.53}$$

The corresponding normal traction is Hertzian, since for similar materials normal and shear loading are uncoupled. Hence, from equations (2.36-2.38),

$$p(x) = \frac{-k}{A} \sqrt{a_1^2 - x^2} \tag{3.54}$$

It will be seen that $p(x)$ and $q(x)$ are proportional and that the initial assumption that no slip takes place is fully justified provided that the coefficient of friction, $f > \tan\alpha$. Should f equal $\tan\alpha$, slip takes place simultaneously everywhere and the contact begins to slide.

The analogous problem of the oblique loading of contacting spheres is presented by Johnson, (1985) and similar results are obtained. Other problems of loading by normal and tangential forces in different sequences were examined by Mindlin and Deresiewicz (1953).

3.5 The effect of bulk stress

Fretting fatigue normally takes place in the presence of bulk stresses within one or both of the contacting bodies arising from some loading other than the contact itself. For example, in a disc of a gas turbine engine, stresses arise because of the centrifugal loading of the disc as it rotates, in addition to the loading imposed by the blades fitted at the circumference. A similar situation can arise in fretting fatigue experiments. Consider, for example, the configuration employed by Hills, Nowell and O'Connor (1988) depicted in fig. 3.13. Here an oscillating bulk stress, σ, is applied to one end of a specimen. This stress causes the specimen to extend and the two cylindrical fretting pads which are clamped to it to move to the right. These pads are restrained by springs so that a tangential fretting force, Q, is applied to the contact in phase with the bulk load. The compliance of the springs may be adjusted to ensure that $Q < fP$ and thus the contact remains in partial slip. This configuration is attractive because it allows application of a fretting force and a bulk stress on a single-actuator testing machine. In analysing the contact conditions occurring it is tempting to assume that the Mindlin-type analysis of §3.2 applies in full and the effect of the bulk stress

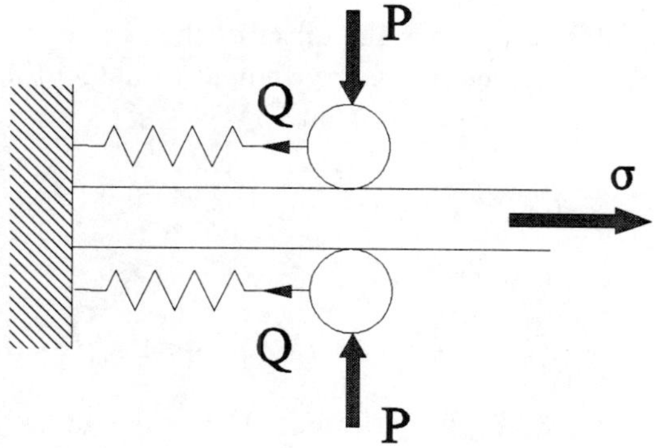

Figure 3.13: Configuration for fretting fatigue experiments: cylindrical pads in contact with a flat specimen.

can be accounted for by superposing a constant tension on the stress field obtained from the contact analysis. However, the results of §3.4 give grounds for caution in applying superposition without careful thought. In this case it will be apparent that the bulk stress, σ, is present in the specimen but not in the pads. This means that the specimen experiences a corresponding bulk strain, ϵ, which is absent from the pads. This mismatch in strains will cause an additional term in the tangential matching equation (3.7) and the resulting shear tractions will differ from those arising in the absence of a bulk stress.

The configuration of fig. 3.13 was first analysed by Nowell and Hills (1987) and will be used here as an example of analysis in the presence of a bulk stress. The tensile stress, σ, will be assumed to be applied to the specimen under conditions of plane strain so that the corresponding strain in the x-direction is given by

$$\epsilon_{xx} = \frac{\sigma(1-\nu^2)}{E} = \frac{\sigma(1-\nu)}{2\mu} \tag{3.55}$$

Let us assume that, of the two bodies in contact, body 2 experiences the bulk tension. The difference in relative tangential displacement gradients ($\frac{\partial g}{\partial x}$) in the stick zone is thus no longer zero but equal to $-\epsilon_{xx}$ i.e.

$$\frac{\partial g}{\partial x} \equiv \frac{\partial u_1}{\partial x} - \frac{\partial u_2}{\partial x} = -\epsilon_{xx} = -\frac{\sigma(1-\nu)}{2\mu} \tag{3.56}$$

Hence equation (3.7) becomes, for similar materials

$$\frac{-\sigma}{4} = \frac{1}{\pi} \int\limits_{contact} \frac{q(\xi)}{(x-\xi)} d\xi \tag{3.57}$$

In contrast to the case without bulk tension, it is unlikely that the stick zone will be centrally positioned and it is therefore necessary to postulate a stick zone extending from $x = e - c$ to $x = e + c$, where e is the offset of the centre of the stick zone from the centre of the contact. Once again it is convenient to model the shear tractions as a perturbation of the full slip solution so that

$$q(x) = fp_0\sqrt{1-(x/a)^2} + q'(x) \tag{3.58}$$

where $q'(x)$ is zero in the slip zones i.e.

$$q'(x) = 0 \qquad |x - e| < c \tag{3.59}$$

Substitution of (3.58) into (3.57), integration of the fully sliding term, and normalisation of the integral gives

$$\int_{-1}^{1} \frac{q'(t)}{s-t} dt = \frac{fp_0 \pi c}{a}\left[s + \frac{e}{c} - \frac{b}{c}\right] \tag{3.60}$$

where $s = (x - e)/c$, $t = (\xi - e)/c$, and $b = \sigma a/(4fp_0)$. Once again this integral equation may be solved using the standard recipes outlined in Chapter 2 (equations (2.23) and (2.32)-(2.34)). A bounded weight function is employed and this gives rise to an additional side (or consistency) condition which must be satisfied (see Nowell, Hills, and Sackfield (1993)). This condition yields, in this case

$$\int_{-1}^{1} \frac{\xi + (e/c) - (b/c)}{\sqrt{1-\xi^2}} d\xi = 0 \tag{3.61}$$

This fixes the value of e so that

$$e = b = \frac{\sigma a}{4fp_0} \tag{3.62}$$

The solution to (3.60) may now be simplified to give

$$q'(x) = -fp_0 \frac{c}{a}\sqrt{1-((x-e)/c)^2} \tag{3.63}$$

It will be noticed that this perturbation of the fully sliding shear tractions is exactly the same as that obtained in §3.2 where no bulk tension was present (3.17) save that the distribution is applied over the shifted stick zone ($|x - e| < c$) of the current problem. Thus, the effect of the bulk tension has merely been to shift the position of the stick zone by an amount e, as shown in fig. 3.14. The sub-surface stress field can be obtained, as before, by superposition of the results for elliptical traction distributions although the shifted origin of $q'(x)$ will now have to be taken into account. One effect of displacing the stick zone is to enhance the in-plane stress component, σ_{xx}, induced at the trailing edge of the contact (i.e. at the edge of the larger slip region, fig. 3.14).

The solution developed above is satisfactory as long as the bulk tension is modest compared with the tangential force. If larger values of tension are applied, one edge of

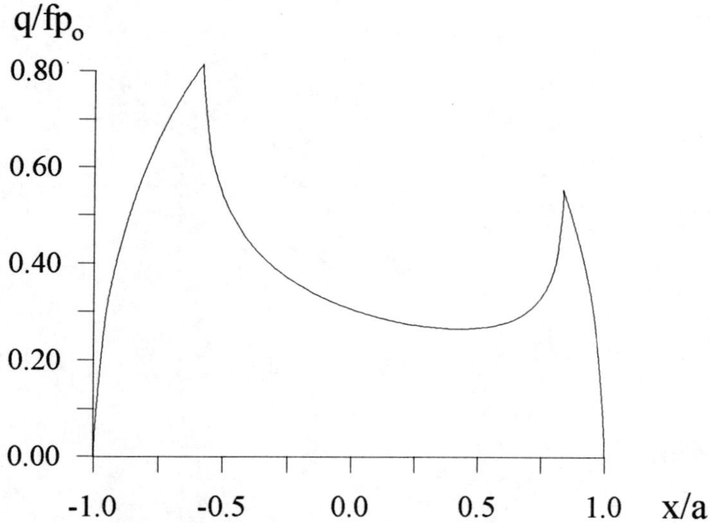

Figure 3.14: Shear traction distribution caused by application of a bulk tension together with a tangential force, $Q/fP = 0.5$, $\sigma/fp_0 = 0.5$.

the stick zone will approach the edge of the contact. The current solution is therefore only valid if $e + c \leq a$, i.e.

$$\frac{\sigma}{fp_0} \leq 4\left(1 - \sqrt{1 - Q/fP}\right) \tag{3.64}$$

If larger values of bulk stress are applied, reverse slip takes place at one edge of the contact (fig. 3.15). In this case an alternative integral equation formulation can be developed. The resulting equation cannot, however, be inverted in closed form and a numerical solution is required (see Nowell and Hills, 1987a). Powerful numerical techniques are available for the solution to this type of Cauchy integral equation (Erdogan, Gupta, and Cook, 1973) and these will be discussed in detail in §4.4.1.

It will be clear from the example discussed above that the presence of a bulk stress in one of the contacting components has a significant effect on the shear traction distribution and thus on the stress field arising from the contact itself. Once again, history-dependence may also be important. If the bulk stress and contact loading are applied in a different (i.e. non-proportional) manner, a different distribution of surface traction will occur. A considerable amount of care is necessary when analysing frictional contacts in partial slip, whether by use of the classical methods for simple geometries presented here or numerical techniques such as finite element analysis.

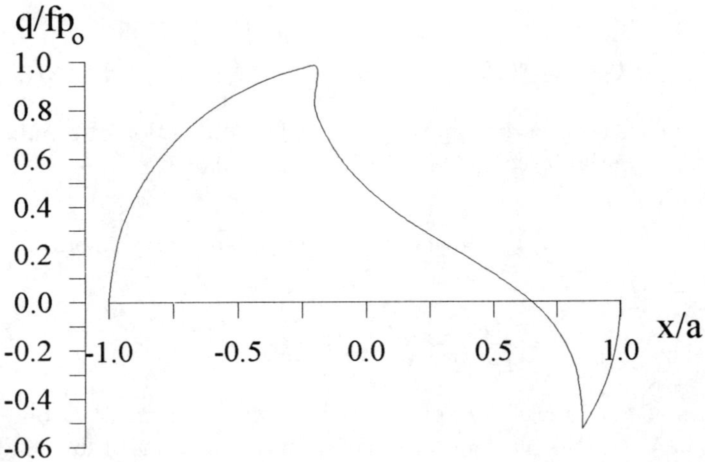

Figure 3.15: Typical shear traction distribution for high values of bulk stress, where reverse slip occurs; $Q/fP = 0.5$, $\sigma/fp_0 = 2.0$.

Chapter 4

Advanced Contact Mechanics

4.1 Introduction

In this chapter a range of more difficult concepts arising in the analysis of contacts will be discussed. This chapter may therefore be omitted at first reading and studied after crack analysis techniques have been mastered. Three specific topics are covered, viz. the influence of elastic material mismatch, contacts subjected to torsion, and the presence of layers in the contacting bodies, including the influence of finite thickness. All of these extensions to the basic theory may be made in closed form, and represent the development of concepts introduced in chapters 2 and 3. There exists, however, a wide range of practically occurring contacts where neither the half-plane nor strip approximations is justified, for example where a contact occurs very close to the edge of a block of material. Under these conditions a completely numerical scheme may be unavoidable. Even where the half-space assumption may be retained, many problems arise which are neither axi-symmetric nor plane - these may profitably be analyzed using a numerical scheme, but based on a half-space influence function formulation, which provides a computationally more efficient strategy than the finite element technique. These numerical approaches to the analysis of contacts are also discussed in the present chapter.

4.2 Contact between elastically dissimilar components

The first question to be addressed is that of contact between elastically dissimilar bodies. In analysing fretting fatigue, one of the primary concerns is the effect of friction, and the final state of stress will therefore depend in detail on the loading history, so that normal contact of the components needs to be dealt with before the influence of an oscillating shearing force can be added. In the elastically similar contacts analysed so far, normal pressure has given rise to identical tangential displacements in both bodies and therefore shear tractions will not arise under purely normal loading. This is not true if the contacting bodies have different elastic properties. Qualitatively, the effect of material mismatch is for the normal tractions (pressure) to give rise to different tangential displacements in the two bodies and conversely for the tangential shear tractions to influence the relative normal surface displacement. It will become apparent, therefore, that mismatch introduces a coupling between the equations for the normal and tangential quantities. Under these circumstances, no analytical solution is possible for anything other than plane or axi-symmetric contacts. A plane contact between cylinders will therefore be considered first, and notes given at the end of the section on the corresponding axi-symmetric problem. As the plane problem itself is invariably very complex only the outline of the solution will be given here. For further details Hills, Nowell and Sackfield (1993) or Nowell, Hills and Sackfield (1988) should be consulted. The underlying governing equations for dealing with plane contact between dissimilar bodies were developed in §2.2, viz. equations (2.16) and (2.21). They are reproduced here for convenience:

$$\frac{1}{A}\frac{\partial h}{\partial x} = \frac{1}{\pi}\int_{-a}^{a}\frac{p(\xi)}{x-\xi}\,d\xi - \beta q(x) \tag{4.1}$$

$$\frac{1}{A}\frac{\partial g}{\partial x} = \frac{1}{\pi}\int_{-a}^{a}\frac{q(\xi)}{x-\xi}\,d\xi + \beta p(x) \tag{4.2}$$

The contact extends between $x = \pm a$, and A, β are defined by equations (2.19) and (2.20). A formal solution will demand traction distributions $p(x), q(x)$ which satisfy both of the above equations, and which simultaneously agree with the stick or slip region requirements discussed extensively in §3.1. This is a very challenging problem, particularly when posed in this form. However, since in most problems the magnitude of the shear tractions is smaller than the normal pressure, the coupling term $q(x)$ in equation (4.1) has a much less significant effect than $p(x)$ in equation (4.2). A common satisfactory approximation which is often made is therefore to neglect the influence of shear traction on the surface normal displacement, so that the contact pressure remains as it was for the corresponding 'similar contact' problem. This simplification was first employed by Goodman (1962) and is often termed the Goodman approximation. Often it provides a good first approximation to the true solution, and may, if desired, be used in the first stage of a sequential iteration procedure in which improved estimates for $p(x)$ and $q(x)$ are obtained by alternating between solutions for normal and shear tractions. It is possible to obtain a closed form solution for the shear traction $q(x)$ with this idealization, starting out with the pressure distribution found from an ordinary Hertz analysis.

The problem to be solved is shown schematically in fig. 4.1. Under normal loading, surface particles tend to displace laterally in proportion to the quantity $(1-\nu)/\mu$ i.e. the body's compliance. Hence, as discussed above, if the two bodies are dissimilar, particles will displace by different amounts, and the relative slip will be resisted by friction. At small values of $|x|$, i.e. near the centre of the contact, the normal pressure

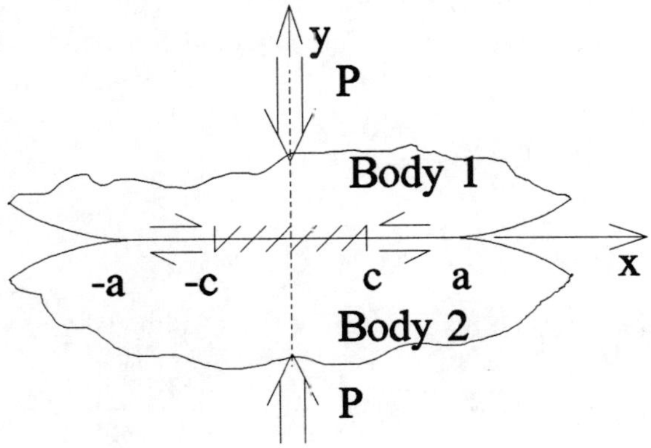

Figure 4.1: Arrangement of stick and slip zones for contact of elastically dissimilar cylinders under the action of a normal force

is sufficient to hold surface particles together, and there is a stick zone. Towards the outside edges of the contact $|x| \rightarrow a$, the contact pressure falls to zero, and hence is insufficient to maintain stick conditions. It is convenient to start by assuming a central stick zone of width $2c$. Within it the relative displacement of surface particles will remain equal to the value at the point of entering the stick zone, as the overall load is gradually increased, and the point in question is enveloped. Furthermore, it is clear from symmetry that the traction distribution in the two slip zones will be equal in magnitude but opposite in sign. These observations may be expressed as follows;

$$
\begin{aligned}
p(x) &= -p_0\sqrt{1-(x/a)^2} \\
\frac{\partial g}{\partial x} &= g_0'(x) \qquad |x| < c \\
q(x) &= fp(x)\mathrm{sgn}(\beta x) \qquad c \leq |x| \leq a
\end{aligned}
\tag{4.3}
$$

where $g_0'(x)$ represents the relative tangential displacement gradient when two contacting particles first enter the stick zone.

Before the integral equation can be solved, the form of function representing relative surface tangential displacement, $g(x)$, must be inferred. This is best done by noting that the only length scale present is the contact size itself. Thus, all contacts between the same pair of materials will be geometrically similar and the stick zone size will be a fixed proportion of the contact half-width, i.e. c/a =constant. This so called self-similarity (Spence, 1973) means that the pre-strain $g_0'(x)$ must be proportional to $|x|$, and hence

$$
g_0'(x) = \frac{C\,|x|}{A} \tag{4.4}
$$

where C is a constant, unknown *a priori*. Substituting equations (4.3) and (4.4) into equation (4.2) and exploiting the anti-symmetry of $q(\xi)$ we find

$$\frac{2}{\pi}\int_0^c \frac{\xi q(\xi)}{x^2-\xi^2}\,d\xi + G(x) = 0 \qquad |x| < c \tag{4.5}$$

where

$$G(x) = \frac{2f}{\pi}\int_c^a \frac{\xi p(\xi)}{x^2-\xi^2}\,d\xi + \beta p(x) - \frac{C\,|x|}{A} \tag{4.6}$$

Integral equation (4.5) may be inverted using the results given in chapter 2 with the result

$$q(x) = \frac{2x}{\pi}\sqrt{c^2-x^2}\int_0^c \frac{G(\xi)}{\sqrt{c^2-\xi^2}(x^2-\xi^2)}\,d\xi \tag{4.7}$$

together with a side condition from the consistency requirements (associated with $q(\xi)$ being bounded at each end of the interval)

$$\int_0^c \frac{G(\xi)}{\sqrt{c^2-\xi^2}}\,d\xi = 0 \tag{4.8}$$

Substituting the definition of $G(\xi)$ (equation 4.6) into the above equation and evaluating the integrals we find, after lengthy algebra, that

$$\beta E(k) = f\left[K'(k) - E'(k)\right] + \frac{Cak}{Ap_0} \tag{4.9}$$

where

$$k = c/a,$$

E and K are the elliptic integrals defined by

$$\begin{matrix} E \\ K \end{matrix}(k) = \int_0^{\pi/2}\left(1 - c^2\sin^2\theta\right)^{\pm\frac{1}{2}}\,d\theta,$$

and the notation $E'(k), K'(k)$ implies replacing k by $\sqrt{1-k^2}$. The value of C was found by Spence (1973), and this enables the stick zone size to be written as the root of

$$\beta K(k) = fK'(k) \tag{4.10}$$

with

$$\frac{C}{A} = \frac{\pi f p_0}{2akK(k)}$$

Lastly, the shear traction distribution itself within the stick zone is given by

$$q(x) = \frac{fp_0}{K(k)}\left\{\sqrt{1-\left(\frac{x}{a}\right)^2}F(k,\phi) - \frac{x}{2c}\ln\left[\frac{c-\sqrt{c^2-x^2}}{c+\sqrt{c^2-x^2}}\right]\right\}\operatorname{sgn}(\beta x) \qquad |x| \le c \tag{4.11}$$

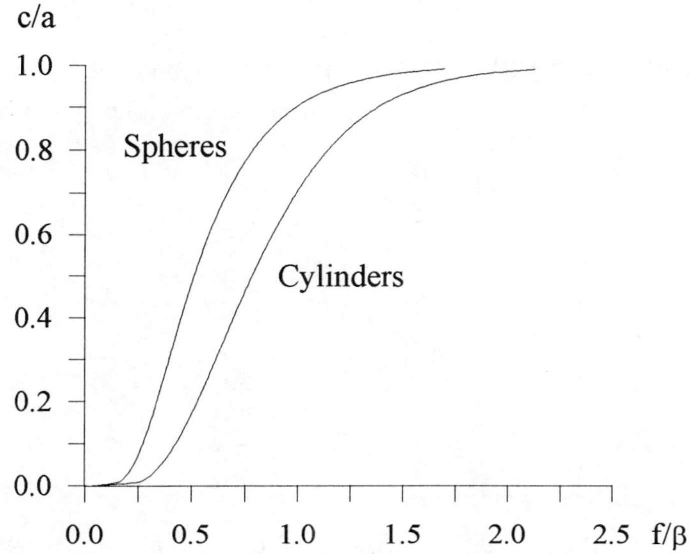

Figure 4.2: Dimensionless size of central stick region, c/a, as a function of the ratio f/β, for normal contact of cylinders, and spheres. The Goodman approximation is employed.

Here, $F(k, \phi)$ is the incomplete elliptic integral of the first kind, given by

$$F(k, \phi) = \int_0^{\phi} \frac{1}{\sqrt{1 - k^2 \sin^2 \theta}} \, d\theta \tag{4.12}$$

The size of the stick zone (equation 4.10) is shown in fig. 4.2, and the distribution of shear traction over the surface is shown in fig. 4.3. It was stated at the outset that the presence of shearing tractions causes different normal surface displacements, and hence modifies the relative curvature of the bodies. In turn this will influence the contact pressure distribution. However, in practice the coupling phenomenon is quite small and will not be addressed here; further details are given in Hills, Nowell and Sackfield (1993).

It is now appropriate to consider the influence of applying a subsequent tangential force. At the end of the normal loading phase, there is an anti-symmetric shear traction distribution, of the form sketched in fig. 4.4(a). As with the case of cyclic tangential loading of similar bodies discussed in §3.4, application of an infinitesimal tangential shear force, ΔQ, causes only a trivial change in the shear traction distribution, but there is a corresponding change in relative tangential displacement of surface particles. In the region $-a < x < -c$ this change violates the requirement that in a slip zone the frictional traction should oppose the slip velocity, and hence the surfaces stick. This gives the stick/slip zone arrangement indicated in fig. 4.4(b). If now a finite, increasing shearing force is imposed, the stick zone $c < x < a$ will

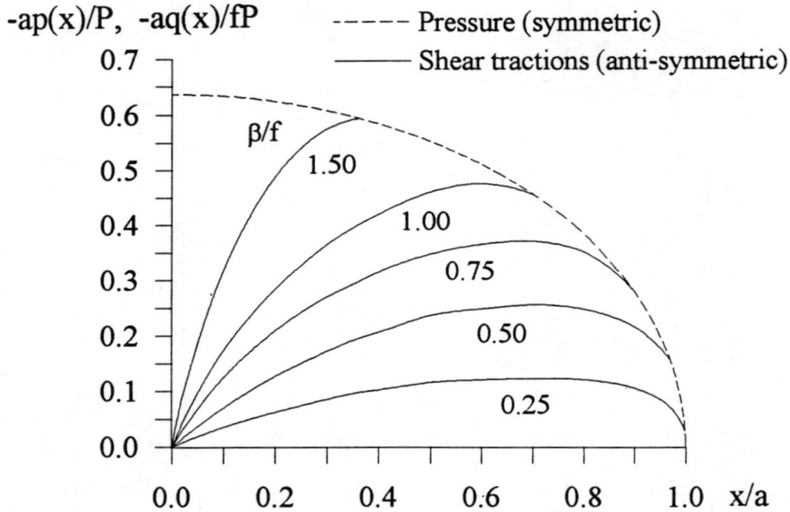

Figure 4.3: Shear traction distribution between elastically dissimilar cylinders.

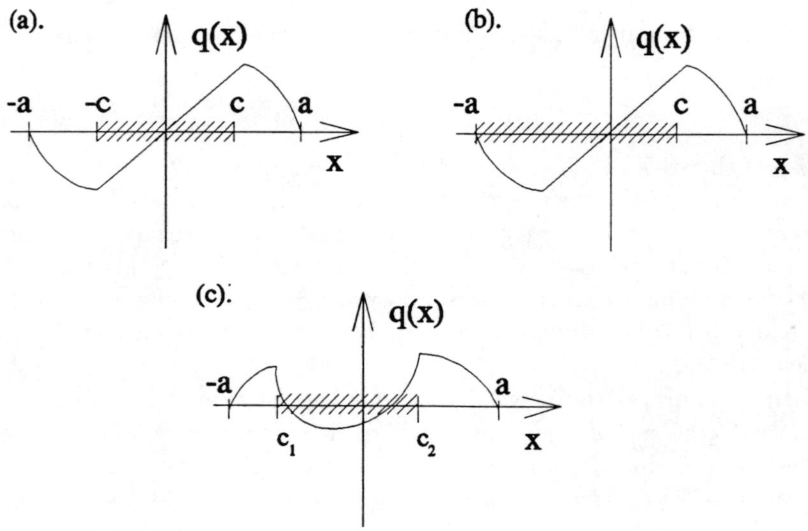

Figure 4.4: Stick/slip arrangement during contact of elastically disimilar cylinders (a) After normal loading alone is applied, (b) After applying an infinitesimal shear force ΔQ, (c) After applying a finite shear force Q.

gradually recede whilst a slip zone of the same sign will start to develop from $x = -a$. The situation is now as depicted in fig. 4.4(c), where adhesion is expected over the range $c_1 < x < c_2$ (both of which are unknown *a priori*) and a slip zone of the same sign exists on either side. To solve for the shear traction distribution, it is convenient to write $q(x)$ as a perturbation, $q_1(x)$, on the fully sliding solution and to make use of equation (4.2). Within the new stick zone the relative tangential strain, $\frac{\partial g}{\partial x}$ is equal to the value obtaining at the end of the normal loading phase, so that

$$\frac{1}{\pi}\int_{\gamma_1}^{\gamma_2} \frac{\hat{q}_1(s)}{r-s}\,ds = -\frac{f}{\pi}\int_{-1}^{1} \frac{\hat{p}(s)}{r-s}\,ds - \beta\hat{p}(r) - \frac{a^2}{AP}\frac{\partial g}{\partial x} \qquad \gamma_1 \le r \le \gamma_2 \tag{4.13}$$

where

$$\begin{aligned} \hat{p}(x) &= ap(x)/P \\ \hat{q}(x) &= aq(x)/P \\ \hat{q}(r) &= f\hat{p}(r) + \hat{q}_1(r) \\ \gamma_1 &= c_1/a \\ \gamma_2 &= c_2/a \\ r &= x/a \\ s &= \xi/a \end{aligned}$$

If we assume that the contact pressure remains Hertzian, the pressure distribution $\hat{p}(x)$ is known explicitly, and hence this is a moderately straightforward integral equation in the unknowns $\hat{q}(x), \gamma_1, \gamma_2$. However, an analytical inversion is clearly impossible as the right hand side of the equation is too complex. Equation (4.13) is a Cauchy singular integral equation, so called because the kernel of the first integral term becomes singular when the integration points, s, and collocation points, r, coincide. Integrals of this kind are difficult to handle, but once again special numerical quadratures may be employed (Erdogan, Gupta and Cook, 1973). The technique will be described in more detail in §4.4.1. In the present case the perturbation on the full-slip shear traction distribution, $\hat{q}_1(r)$, is required to vanish as the edges of the stick zone are approached. Hence $\hat{q}_1$ is written as the product of a fundamental function with the correct end-point behaviour and a bounded component, ϕ, i.e.

$$\hat{q}_1(s) = \sqrt{(s-\gamma_1)(\gamma_2-s)}\,\phi(s) \tag{4.14}$$

so that $\phi(s)$ is a well-behaved function. Equation (4.13) may now be discretised and becomes

$$\sum_{i=1}^{n} \frac{1-\lambda_i^2}{n+1}\frac{\phi(\lambda_i)}{r_k - s_i} = -\frac{f}{\pi}\int_{-1}^{1}\frac{\hat{p}(u)}{r_k-u}\,du - \beta\hat{p}(r_k) - \frac{a^2}{AP}\frac{\partial g(r_k)}{\partial x} \qquad k = 1, 2, \ldots n+1 \tag{4.15}$$

where

$$
\begin{aligned}
r_k &= B + \eta_k C \\
s_i &= B + \lambda_i C \\
B &= \tfrac{1}{2}(\gamma_1 + \gamma_2) \\
C &= \tfrac{1}{2}(\gamma_1 - \gamma_2) \\
\lambda_i &= \cos\left(\tfrac{i\pi}{n+1}\right) \qquad i = 1, 2, ...n \\
\eta_k &= \cos\left(\tfrac{(2k-1)\pi}{2(n+1)}\right) \qquad k = 1, 2, ...n+1
\end{aligned}
$$

It will be noted that this scheme yields $n+1$ equations to determine the n unknowns $\phi(\lambda_i)$. The additional equation is needed to ensure that γ_1 and γ_2 are consistently chosen; in practice values of both γ_1 and γ_2 are guessed and one equation is dropped from the set implied by (4.15). This set is then solved and the extra equation checked to see if it is satisfied. If it is not, a new guess for γ_2 is made and the process is repeated. Results from the above calculation are shown in fig. 4.5. These give a comparison of the traction distribution and stick zone position for a monotonically increasing shear force, Q. The equivalent solution given by the Mindlin calculation is also provided for comparison. The value of β chosen is 0.4 which is the maximum which may in practice be expected, and the coefficient of friction is taken as 0.5. The process described above may be extended to cover the case when the shearing force is cycled. Results for a typical case in the steady state are depicted in fig. 4.6.

The plane contact between dissimilar bodies described above affords an almost analytical solution, save for the last phase of solving the integral equation. The solution is useful insofar as it indicates how the shear traction distribution is perturbed. However, the assumption of plane deformation limits the usefulness of the solution in relation to test geometries, as edge effects will still be present. As mentioned in §3.3, when designing an experiment to test fretting fatigue strength under partial slip conditions, a very well defined contact may be obtained by using spherical pads pressed onto the main test specimen. However, the number of materials available in the form of spheres is somewhat limited, although the excellent surface finish commonly found on commercial steel balls means that they may be a better choice than specially manufactured spherical pads. Invariably, therefore, the contact will be between dissimilar materials unless the test material is steel, and the problem of partial-slip (or 'Mindlin contact') between dissimilar spheres must be addressed. The Mindlin-Cattaneo technique for analysing partial slip has already been described in Chapter 3, and the discrepancy associated with the presence of a transverse displacement component pointed out. This discrepancy may be increased when the contacting bodies are elastically dissimilar, and a full solution to the exact problem demands a numerical solution (Munisamy, Hills and Nowell, 1992), which will be summarised here. The basic geometry of the problem is shown in fig. 4.7. At the outset the Goodman approximation will be made, i.e. that the shear tractions present do not affect the normal pressure, as described earlier in this section in relation to the associated plane problem. The principal task is therefore to determine how the contact patch is split into regions of stick and regions of slip, and to ensure that the following conditions

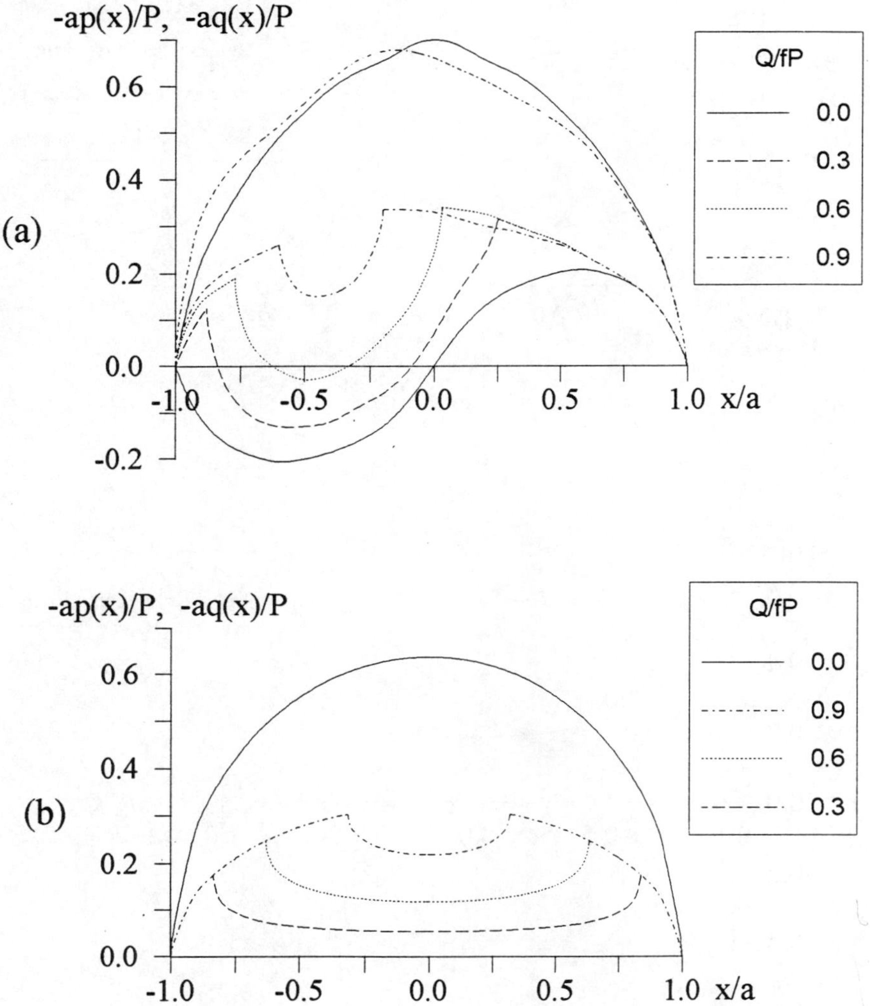

Figure 4.5: Traction distribution during tangential loading of elastically dissimilar cylinders by a monotonically increasing shear force. (a) $\beta = 0.4, f = 0.5$, solution using the scheme described. (b) Mindlin solution, i.e. $\beta = 0$.

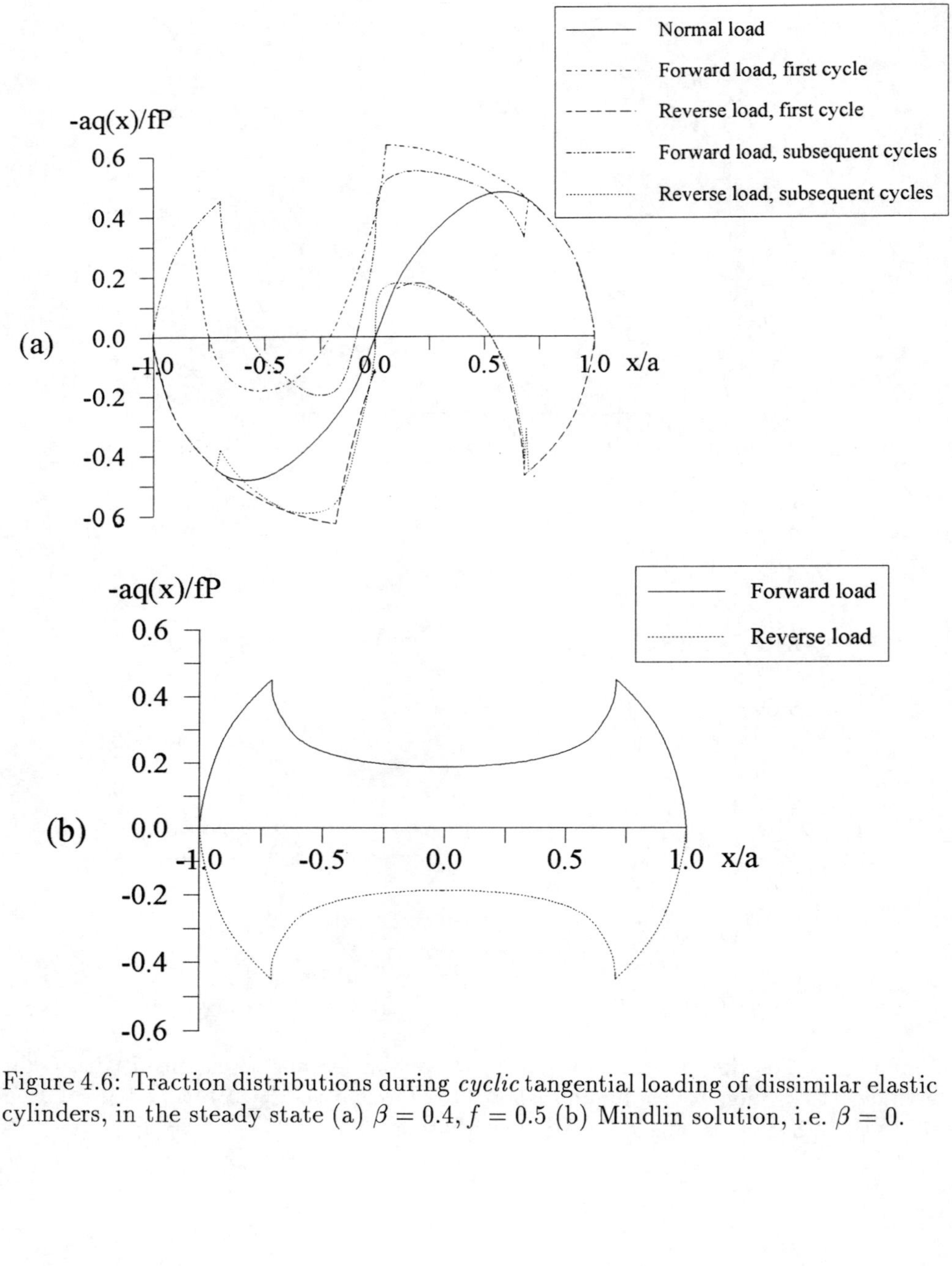

Figure 4.6: Traction distributions during *cyclic* tangential loading of dissimilar elastic cylinders, in the steady state (a) $\beta = 0.4, f = 0.5$ (b) Mindlin solution, i.e. $\beta = 0$.

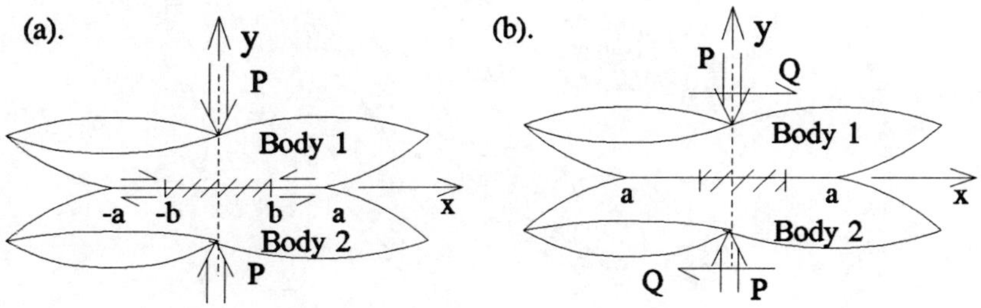

Figure 4.7: Geometry of the problem of partial slip between two elastically dissimilar spheres (a) After normal loading alone, (b) When a shearing force, Q, is applied. The stick zone is now of arbitrary shape.

are satisfied within each zone:

(a) Stick Region First, the shear traction $\mathbf{q}(x, y)$, which may have any orientation in the x, y plane (and is therefore expressed now as a vector quantity), must be less than the friction-limited value:

$$|\mathbf{q}(x,y)| < -fp(x,y) \tag{4.16}$$

Secondly, if the applied shearing force Q is increasing monotonically, so that the stick region is diminishing, the relative incremental displacements, s_x, s_y must be zero:

$$\frac{ds_x}{dQ} = \frac{ds_y}{dQ} = 0 \tag{4.17}$$

where

$$s_x(x,y) = \Delta_x - (u_{x1} - u_{x2})$$
$$s_y(x,y) = -(u_{y1} - u_{y2})$$

Here Δ_x is the relative displacement in the x-direction of two points deep within the bodies and u_{ij} is the local surface elastic deformation of body j in direction i.

(b) Slip Region Within the slip region the shear traction takes its limiting value, i.e.

$$|\mathbf{q}(x,y)| = -fp(x,y) \tag{4.18}$$

Furthermore, the direction of slip must always be such as to oppose the shear traction. This requirement is expressed by the condition

$$\frac{\mathbf{q}(x,y)}{|\mathbf{q}(x,y)|} = -\frac{d}{dQ}\left[\frac{\mathbf{s}(x,y)}{|\mathbf{s}(x,y)|}\right] \tag{4.19}$$

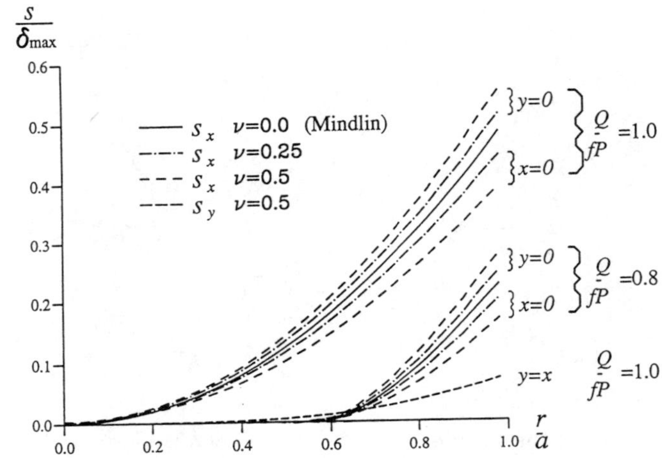

Figure 4.8: Components of relative slip between elastically similar bodies: the influence of Poisson's ratio (from Munisamy, Hills, and Nowell (1993) with the permission of ASME).

Lastly, the overall equilibrium condition requires that

$$\int_{\text{contact}} q_x(x, y) = Q \tag{4.20}$$

The solution cannot be obtained analytically. The contact region is instead split into small rectangular elements which may be either adhering or slipping (Munisamy, Hills, and Nowell, 1992), and the appropriate inequalities derived above are satisfied. Here, discussion will be restricted to a consideration of the results. As an introduction we shall examine briefly the case when the contacting bodies are elastically similar. The Mindlin solution (§3.3) is then *exact* if the material shows no Poisson effect ($v = 0$), so that $s_y \equiv 0$. For other cases, the numerical results are summarized in fig. 4.8. The Mindlin solution retains axi-symmetry, and therefore the estimated value of s_x is the same along all radial lines. The calculated values of s_x differ from this value, and the discrepancy increases with increased Poisson's effect. Furthermore, the discrepancy is positive along a line perpendicular to the direction of shearing (i.e. along $x = 0$) but negative along the line of shearing ($y = 0$). The maximum value of transverse shear, s_y, occurs along the lines $x = \pm y$ and is included in fig. 4.8. Another way of emphasising the difference between the Mindlin solution and an exact solution is to examine the surface density of frictional energy dissipation, i.e. the product $\mathbf{s}.\mathbf{q}$ at each point. This parameter has particular significance in the analysis of fretting fatigue (see Chapter 9). In the Mindlin solution $\mathbf{s} \equiv s_x$ and $\mathbf{q} \equiv q_x$; each varies only with radial position but not angular position so that a contour plot of energy dissipation over the contact would appear as a set of concentric rings. Figure 4.9 shows the corresponding solution for incipient sliding conditions and an incompressible material. It is readily

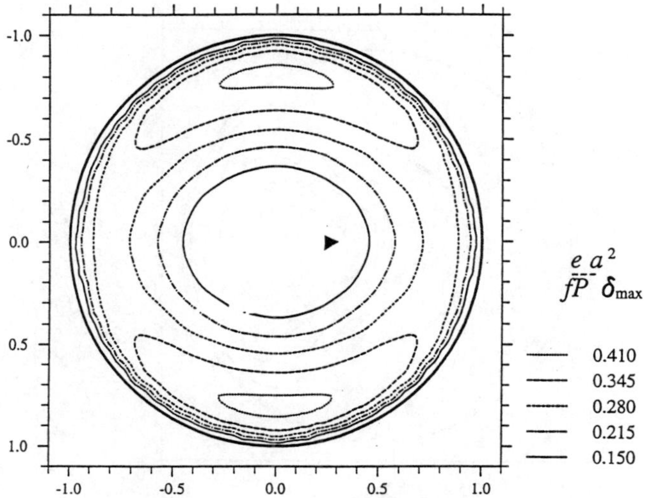

Figure 4.9: Surface density of frictional energy dissipation for elastically similar contact between incompressible bodies ($\nu = 0.5$) and incipient sliding ($Q/fP = 1.0$).

seen that the dissipated energy is localised on either side of the plane of symmetry ($y = 0$) and that the maxima are of the order of 25% greater than the value predicted on the plane of symmetry. Hence, the localization of frictional energy expenditure is significant; however, the overall expenditure, as characterised by a force-displacement hysteresis loop is imperceptibly different.

For the case of dissimilar materials, we note that, in order to formulate the problem properly, we should take the stick/slip regime corresponding to normal indentation as the starting point. This consists of a central disc of adhesion (Spence, 1975) surrounded by a annulus of radial shearing traction. The size of the stick zone is given by

$$\frac{\beta}{f} = \frac{kK(k')}{\ln\left(\frac{1+k}{1-k}\right)} \tag{4.21}$$

where $k = c/a$ is the dimensionless stick zone radius and K is an elliptic integral of the first kind. The shear traction in the slip annulus is given in Hills and Sackfield (1987). Values of the stick zone size are included in fig. 4.2 for comparison with the equivalent two-dimensional problem. As an example, we consider a pair of materials where $\beta/f = 1.0$, $k = 0.45$, as indicated by the broken circle on fig. 4.10. Application of an infinitesimal shearing force Q then causes adhesion over the region shown by the chain line. As the applied load is increased further, the zone of adhesion recedes, but to an off-centre location as depicted. The presence of the pre-existing shear stress associated with normal indentation introduces additional tangential compliance and produces an offset stick zone in the steady state (i.e. after several cycles of loading).

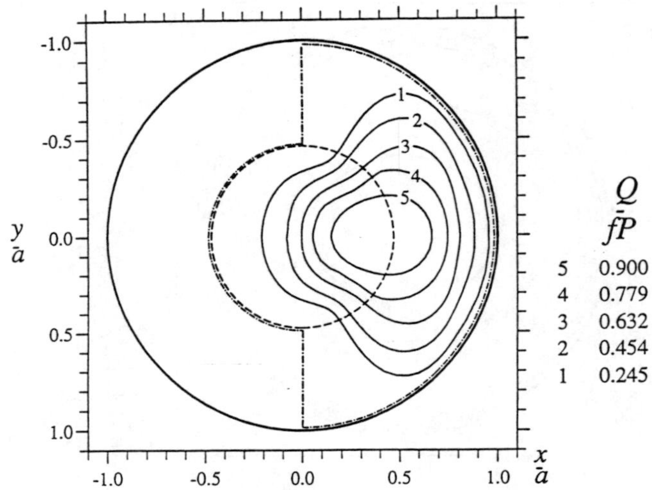

Figure 4.10: Arrangement of stick/slip regions for a dissimilar contact when $\beta/f = 1.0$. The broken line encloses the stick zone after normal indentation. The chain line indicates the stick zone after the imposition of an infinitesimal shear.

Thus, although the force-displacement hysteresis loop is displaced, the area enclosed (i.e. the energy dissipated) is close to the Mindlin solution. As before, the frictional dissipation is localized in two crescent-shapes on either side of the plane of symmetry.

4.3 Twisting contacts

In each of the contacts studied so far the varying force has been applied tangentially and in the plane of the contact. Another form of loading which may be exerted on a notionally static arrangement is a torque about the normal to the contact disc. This gives rise to a well-defined partial slip problem which has not hitherto gained much attention in the context of fretting fatigue although the geometry is attractive from an experimental point of view and may well simulate certain practical situations such as torsional loading of a bolted connection or inter-strand contact in wire ropes under tension (Raoof, 1993). Here the analysis of two contacting spheres subjected to a twisting torque will be presented. It will be assumed that the two spheres have been pressed together to form a conventional axi-symmetric Hertzian contact over a disk of radius a, fig. 4.11. The two bodies will initially adhere over the entire contact (if they are elastically similar), but the application of a monotonically increasing torque T might be expected to produce an annulus of slip, (inner radius c) which migrates inwards. Within the slip annulus the shear stress will be limited by friction, whilst within the central adhesion zone particles will move tangentially as a rigid body, and

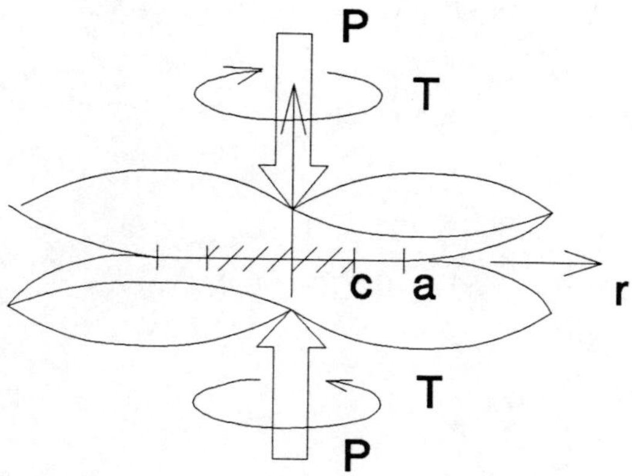

Figure 4.11: Two spheres pressed together by a normal force, P and subjected to a subsequent twisting torque, T.

hence

$$\tau_{z\theta} = f p_0 \sqrt{1-(r/a)^2} \qquad c \leq r \leq a \tag{4.22}$$

$$u_\theta = \theta r \qquad 0 \leq r \leq c \tag{4.23}$$

The problem of finding a shear stress distribution $\tau_{z\theta}(r)$, $(0 < r < c)$, which, together with the traction within the slip annulus produces a surface displacement given by equation (4.23) was solved by Lubkin (1951). He showed that the traction distribution required is given by

$$\frac{\tau_{z\theta}}{f p_0} = \sqrt{1-\left(\frac{r}{a}\right)^2}\left[1+\frac{2}{\pi}\left\{k'^2 D'(k) F(k,\phi) - K'(k) E(k,\phi)\right\}\right] \qquad 0 \leq r \leq c \tag{4.24}$$

where

$$\begin{aligned} \sin\phi &= \tfrac{1}{k}\sqrt{\tfrac{c^2-r^2}{a^2-r^2}} \\ k &= c/a \\ k'^2 &= 1-k^2 \\ D'(k) &= (K'(k)-E'(k))/k'^2 \\ K'(k) &= K(k') \\ E'(k) &= E(k') \end{aligned}$$

$K(k), E(k)$ are complete elliptic integrals of the first and second kind respectively, and $F(k,\phi), E(k,\phi)$ are the corresponding incomplete integrals. The surface traction distribution associated with a monotonically increasing torque is shown in fig. 4.12. An analysis of the complete subsurface stress state was given by Hills and Sackfield

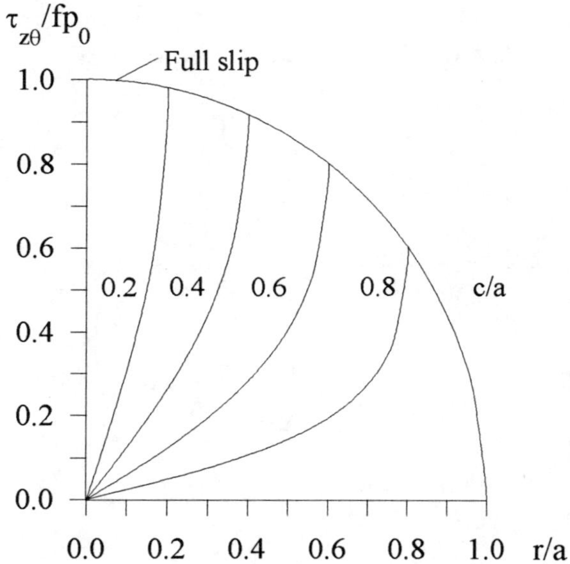

Figure 4.12: Shear traction distribution for a contact undergoing twist.

(1986). The angle of twist, θ_i, each body undergoes is given by [1]

$$\frac{4\pi\mu_i a^2\theta_i}{3fP} = \frac{2\mu_i\theta_i}{fp_0} = k^2 D'(k) \qquad i = 1, 2 \tag{4.25}$$

where μ_i is the modulus of rigidity of body i. The torque required to achieve this relative rotation may be found by ensuring overall equilibrium, i.e.

$$T = 2\pi \int_0^a \tau_{z\theta}(r) r^2 dr \tag{4.26}$$

However, this integral may not be evaluated in closed form. (see Lubkin (1951) or Hills, Nowell and Sackfield (1993) for further discussion of this point). If the slip annulus is small the following asymptotic expressions produced by Deresiewicz (1954) are useful: first, the size of the slip annulus is given approximately by

$$\frac{T}{fPa} \simeq k'^2 \left(1 - \frac{3}{8}k'^2 - \frac{1}{64}k'^4 - ....\right) \tag{4.27}$$

The error involved in truncating to three terms is less than 0.5% if $k > 0.7$. Secondly the equation relating twist to the size of the annulus (4.25) may be rewritten avoiding

[1] Note that the total relative angle of twist between the two bodies has two contributions of the kind given by (4.25), one for each body. If the two bodies are elastically dissimilar, (4.25) will hold only approximately for each body since the existence of a pre-existing radial shearing traction τ_{rz} has been ignored.

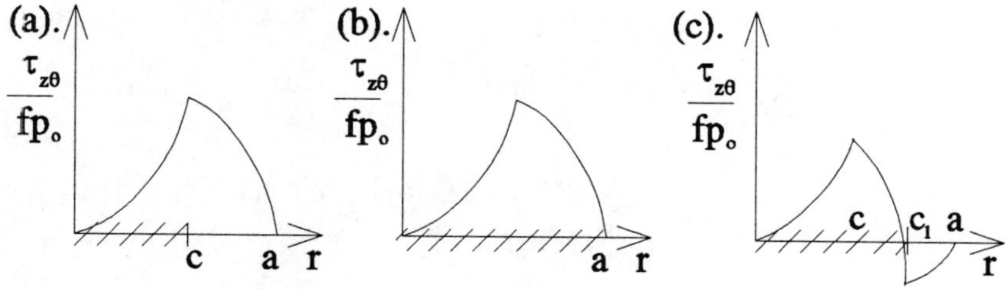

Figure 4.13: Shear traction distribution under an oscillating torque: (a) traction state after loading to the peak value, (b) effect of an infintesimal reduction in torque, giving rise to adhesion everywhere, (c) effect of reducing torque by a finite amount giving an annulus of reverse slip.

elliptic integrals in the form

$$\frac{\mu_i a^2}{fP} \simeq \frac{3}{16} k'^2 \left(1 + \frac{3}{8} k'^2 + \frac{15}{64} k'^4 + ...\right) \tag{4.28}$$

Here, the error is less than 2.9% if $k > 0.7$. Equations (4.27) and (4.28) may be combined to produce a very approximate torque-twist relation if $T \ll fPa$

$$\frac{\mu_i a^2}{fP} \simeq \frac{1}{8} \left[1 - \sqrt{1 - \frac{3T}{2fPa}}\right] \left[3 - \sqrt{1 - \frac{3T}{2fPa}}\right] \tag{4.29}$$

The above analysis concerns the traction distribution occurring under a monotonically increasing torque. The case of cyclic loading has been studied by Deresiewicz (1954), and will now be summarised. It is assumed that a steadily increasing torque has been applied, up to a maximum value T^*; the corresponding radius of the stick zone disc is c, fig. 4.13(a). The torque is now reduced infinitesimally. This results in a negligible change in the traction, but the change is relative surface displacement in what was the slip annulus is directed in the opposite sense, giving rise to adhesion over the entire contact, fig. 4.13(b). A further reduction in torque to a value T now permits backslip over a thin annulus ($c_1 < r < a$, fig. 4.13(c)), the shear traction distribution being formed by superposition. It is straightforward to determine that the result is

$$\begin{aligned}
\tau_{z\theta} &= -fp_0\sqrt{1-\left(\tfrac{r}{a}\right)^2} && c_1 < r < a \\
&= -fp_0\sqrt{1-\left(\tfrac{r}{a}\right)^2}\left[1+\tfrac{4}{\pi}\left\{k_1'^2 D'(k_1)F(k_1,\phi_1) - K'(k_1)E(k_1,\phi_1)\right\}\right] && c < r < c_1 \\
&= -fp_0\sqrt{1-\left(\tfrac{r}{a}\right)^2}\left[1+\tfrac{4}{\pi}\left\{k_1'^2 D'(k_1)F(k_1,\phi_1) - K'(k_1)E(k_1,\phi_1)\right\}\right. \\
&\qquad \left. -\tfrac{2}{\pi}\left\{k'^2 D'(k)F(k,\phi) - K'(k)E(k,\phi)\right\}\right] && 0 < r < c
\end{aligned} \tag{4.30}$$

The quantities k_1, k_1', ϕ_1 are defined in a similar way to k, k', ϕ (see equation 4.24) except that the quantity c in the argument is replaced by c_1. The torque sustained during unloading is found by ensuring equilibrium with the shearing tractions (equation 4.26) but a precise closed form equation is unobtainable. Approximately, the stick zone radius, c_1, during unloading is related to the current torque, T, by

$$\frac{c_1}{a} = \frac{1}{\sqrt{3}}\sqrt{4\left(1-\frac{3(T^*-T)^{\frac{1}{2}}}{4fPa}\right)-1} \tag{4.31}$$

Deresiewicz (1954) also determined the energy lost per cycle for elastically similar spheres under steady state oscillatory torsional loading and found it to be

$$E = \frac{2f^2P^2}{\mu a}\left\{\frac{8}{9}\left[1-\left(1-\frac{3T^*}{2fPa}\right)^{\frac{3}{2}}\right] - \frac{T^*}{fPa}\left[1+\left(1-\frac{3T^*}{2fPa}\right)^{\frac{1}{2}}\right]\right\} \tag{4.32}$$

If the torque is small (i.e. $T^* \ll fPa$) this equation simplifies to

$$E = \frac{3T^{*3}}{16\mu f P a^4} \tag{4.33}$$

Torsional loading of contacting bodies giving rise to a Hertzian pressure distribution over an ellipse has been studied by Mindlin (1949) for the case of an adhered contact. A full solution including the presence of slip was provided by Pacelli (1956).

4.4 Numerical methods: homogeneous bodies

Although closed-form solutions for elastic contacts form an attractive basis for the analysis of carefully controlled fretting fatigue experiments, they have limited applicability in treating real engineering components. In this section some of the assumptions made at the outset in Chapter 2 will be weakened, so that the repertoire of solutions is increased. It is inevitable, however, that the more assumptions are relaxed the more numerically intensive the ensuing calculation becomes. To begin with the assumption that the contacting bodies may still be represented by half-spaces will be retained, so that there are no adjacent boundaries which will weaken the components in the neighbourhood of the contact. However, in contrast to the problems studied in Chapter 2 it will be taken that the bodies have a complex profile so that a closed form integral equation solution cannot be found. A number of possible approaches may be used:

1. The conventional integral equation formulation may be employed, but if the resulting equation cannot be inverted analytically, a numerical solution technique may be employed. Such a method has already been presented in outline form in §4.2.

2. Other types of contact, particularly those which are three dimensional, are conveniently analysed by splitting the contact area up into finite regions over which a prescribed traction element is taken to act. The magnitudes of these elements are then adjusted in order to satisfy the appropriate boundary conditions. This approach is often referred to as the influence function method.

3. The third possible approach is to use modified versions of general stress analysis techniques, such as finite or boundary element methods. However these are often cumbersome when applied to contact analysis and other techniques are to be preferred. Nevertheless they are often the only feasible approach if it is not possible to represent the contacting bodies as half-planes (e.g. in a conformal contact such as that of a pin in a hole).

These three numerical approaches will be examined in turn.

4.4.1 Numerical solution of integral equations

It is often possible to formulate a two-dimensional contact problem in terms of a Cauchy singular integral equation. The basic approach was covered in §2.2. For problems involving elastically similar materials, integral equations of the form given by (2.22) are obtained in both the normal pressure and shear tractions, e.g.

$$\frac{1}{\pi}\int_{-a}^{a}\frac{p(\xi)}{x-\xi}\,d\xi=\frac{1}{A}\frac{\partial h}{\partial x}\qquad |x|\leq a \tag{4.34}$$

For simple functions on the right hand side, equations of this type may be inverted analytically using the recipes given in Chapter 2. However, more complicated functions require the use of numerical quadratures and a particularly powerful technique is described by Erdogan, Gupta, and Cook (1973). A brief summary of the method will be presented here.

		Behaviour at $r = 1$	
		Bounded	Singular
Behaviour at $r = -1$	Bounded	$w(r) = (1 - r^2)^{\frac{1}{2}}$	$w(r) = (1 + r)^{\frac{1}{2}}(1 - r)^{-\frac{1}{2}}$
	Singular	$w(r) = (1 + r)^{-\frac{1}{2}}(1 - r)^{\frac{1}{2}}$	$w(r) = (1 - r^2)^{-\frac{1}{2}}$

Table 4.1: Appropriate weight functions for different end point behaviour

The first step is to normalise the interval of integration by appropriate change of variables to produce an equation of the form

$$\frac{1}{\pi}\int_{-1}^{1}\frac{p(r)}{s-r}\,dr = R(s) \qquad |s| \leq 1 \tag{4.35}$$

The unknown function, $p(r)$, is represented as the product of a bounded function, $\phi(r)$, and a fundamental function, $w(r)$, which takes into account the underlying behaviour expected in the solution. Two possible types of behaviour exist at each end of the interval

1. The unknown function falls continuously to zero, as occurs, for example, with the pressure distribution in 'incomplete contacts' (see Chapter 2).
2. The unknown function is singular, such as occurs in 'complete' contacts, where one of the contacting bodies has a sharp corner.

Four types of fundamental function are therefore possible, according to which behaviour is adopted at each end of the interval. Appropriate choices of fundamental function are summarised in Table 4.1. As an example the case of bounded behaviour at both ends of the interval will be presented here. Analysis of the other cases is broadly similar and can be found in Erdogan, Gupta and Cook (1973) or Hills, Nowell and Sackfield (1993). Equation (4.35) now becomes

$$\frac{1}{\pi}\int_{-1}^{1}\frac{\phi(r)\sqrt{1-r^2}}{s-r}\,dr = R(s) \qquad |s| \leq 1 \tag{4.36}$$

A Gauss-Chebyshev integration scheme is used to represent the integral in equation (4.36). In effect this means that the unknown function, $\phi(r)$, is represented by a series of Chebyshev polynomials. The equality in (4.36) is enforced at a number of collocation points, s_k, and this enables the multipliers of the polynomials to be determined. Thus, the integral equation (4.36) is replaced by N simultaneous linear algebraic equations, where N may be chosen by the user to give an appropriate degree of refinement. For the case being considered the quadrature yields

$$\frac{1}{N+1}\sum_{i=1}^{N}\frac{\phi(r_i)(1-r_i^2)}{r_i - s_k} = R(s_k) \tag{4.37}$$

where the integration points r_i are given by

$$r_i = \cos\left(\frac{i\pi}{N+1}\right) \qquad i = 1, 2, ...N$$

and the collocation points s_k by

$$s_k = \cos\left(\frac{(2k-1)\pi}{2(N+1)}\right) \qquad k = 1, 2, ...N+1$$

At first sight the problem appears overconstrained since there are $N+1$ equations for the N unknowns, $\phi(r_i)$. However, Erdogan, Gupta and Cook (1973) show that the consistency condition required for bounded-bounded behaviour (see Chapter 2) is automatically satisfied by the choice of fundamental function so that one of the equations may be dropped and a solution obtained using any suitable technique for solving simultaneous linear equations (e.g. Gaussian elimination or LU decomposition). Once a solution has been obtained it may be demonstrated that the $N + 1$th equation is indeed satisfied. Values of the original variable, $p(r)$, may be recovered at the integration points from

$$p(r_i) = \phi(r_i)\sqrt{1 - r_i^2} \tag{4.38}$$

The integral of $p(r)$ may also be expressed in discretised form so that

$$\int_{-1}^{1} p(r)\, dr = \frac{\pi}{N+1}\sum_{i=1}^{N}(1 - r_i^2)\phi(r_i) \tag{4.39}$$

If the unknown variable is normal or shear traction this provides a convenient means of evaluating the total load. This method of discretising Cauchy singular integral equations is an extremely efficient and accurate means of solution. It will be employed again in Chapter 6 in connection with the analysis of cracks. An alternative form of quadrature, based on the results of Abramowitz and Stegun (1972), is presented by Gladwell (1980) as part of a review of techniques for the solution of integral equations.

4.4.2 Influence function methods

The integral equation technique in the form described above is not, unfortunately, suitable for three-dimensional problems and an alternative method of linearising the problem must be sought. One possibility is to divide the contact area into a series of finite domains and to model the tractions in each domain in some prescribed manner, for example as uniform or linear over the region. The magnitudes of these traction elements can then be adjusted in order to satisfy the boundary conditions of the problem. The technique relies on establishing the influence of the chosen type of traction element on the displacement at some remote point, and is therefore referred

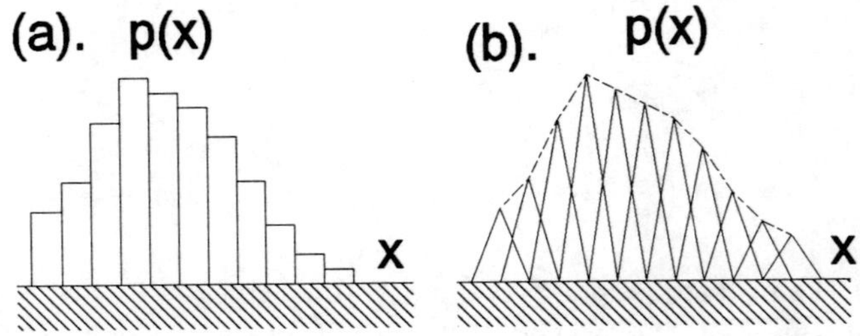

Figure 4.14: Influence function principles applied to plane contacts; (a) piecewise step, (b) piecewise linear.

to as the influence function method. Before proceeding to discuss three-dimensional configurations it is instructive to develop a method for plane problems which may then be extended to deal with three-dimensional situations. Two types of traction distribution will be considered here, and these are shown in fig. 4.14. In the first approximation, fig. 4.14(a) the traction is represented in a piecewise step fashion. To analyse a problem involving normal pressure only the contact patch is split into N elements of width $2c = 2a/N$, and, for a general element i the pressure is denoted by p_i. The value of the surface normal displacement at some other element, j, may then be written as

$$v_j = g_{ij} p_i \tag{4.40}$$

where the quantity g_{ij} is known as the influence function. Because, for two dimensional problems, it is not possible to calculate the absolute value of displacement (see §2.2) it is convenient to calculate v_j as the relative displacement of element j to that produced at some origin which will be kept constant for all elements. When a complete set of elements $i = 1, 2, ..N$ has been imposed, the total displacement at the jth point is given by

$$v_j = \sum_{i=1}^{N} g_{ij} p_i \tag{4.41}$$

In practice the values of the pressure, p_i, are unknown and the surface displacement is that imposed by the contacting body which, for the time being, will be assumed to be perfectly rigid. If the profile of the contacting body is $h(x)$, the displacement at the centre of each pressure element must be matched to $h(x_j)$. Note that this procedure requires the assumption of a particular contact dimension, before solving for the contact pressure idealisation. The total contact load, P, may be found from the solution by ensuring overall equilibrium, i.e.

$$P = -\int p(\xi)\, d\xi = 2c \sum_{i=1}^{N} p_i \tag{4.42}$$

In practice the contact load is often known whilst the contact dimension is to be deduced; it is very difficult to solve the problem posed in this form directly and a satisfactory procedure is to iterate on the contact patch size until the desired contact load is obtained. In order to implement the procedure, the influence function g_{ij} is required. For constant pressure p_i applied over a region $|x| < c$ the normal displacement at a point x_j is given by

$$v_j = \frac{-Ap_i}{2\pi}\left[(x_j + c)\ln\left(\frac{x_j + c}{c}\right)^2 - (x_j - c)\ln\left(\frac{x_j - c}{c}\right)^2 + \text{constant}\right] \qquad |x| \geq c \tag{4.43}$$

Here the compliance, A, is given by

$$A = \frac{1-\nu}{\mu} = \frac{2(1-\nu^2)}{E} \tag{4.44}$$

where ν is Poisson's ratio, μ the modulus of rigidity and E Young's modulus. The required influence function g_{ij} follows immediately by setting $x_j = 2(j - i)c$.

The principal difficulty with using constant pressure elements, as described above, is that they do not provide a very satisfactory means of modelling the contact pressure where it falls continuously to zero, as it does, for example, at edges of an incomplete contact. A better method of representing such behaviour is to use an influence function based on a triangle of traction, as shown in fig. 4.14(b). If the elements are made to overlap a piecewise-linear modelling of the contact tractions is effected. As this readily achieves the need to simulate a contact pressure which falls continuously to zero it is much to be preferred. The same general principles as those developed above in connection with a constant pressure element may be used. The normal displacements at $x = x_j$ for a triangle of maximum pressure p_i applied over the region $|x| < c$ provide the necessary influence functions and are given by

$$v_j = \frac{-Ap_i}{4\pi c}\left[(c + x_j)^2\ln\left(\frac{c + x_j}{c}\right)^2 - (c - x_j)^2\ln\left(\frac{c - x_j}{c}\right)^2 - 2x_j^2\ln\left(\frac{x_j}{c}\right)^2 + \text{const}\right] \tag{4.45}$$

for $|x_j| < c$ and

$$v_j = \frac{-Ap_i}{4\pi c}\left[(c + x_j)^2\ln\left(\frac{c + x_j}{c}\right)^2 - (x_j - c)^2\ln\left(\frac{x_j - c}{c}\right)^2 - 2x_j^2\ln\left(\frac{x_j}{c}\right)^2 + \text{const}\right] \tag{4.46}$$

for $|x_j| > c$.

The formulation allows contact problems to be expressed as a set of linear equations in the unknown magnitudes of the traction elements, p_i. These may be solved directly using a computer library routine or using a variational approach as described by Kalker (1990). A numerical difficulty which arises in solving problems by this approach is that if the 'mesh' of elements is refined too far, the matrix of influence coefficients becomes rather stiff, and difficult to invert. This is easy to appreciate on physical grounds as it is clear that adjacent rows of the matrix will be very similar, and hence the corresponding determinant is rather small. In order to solve problems involving shear traction, the corresponding influence functions for elements of shear traction are also required. Equations (2.12) and (2.13) reveal that for loading of a half-plane the tangential displacement caused by tangential loading is equal in magnitude to the normal displacement caused by a corresponding normal load. Thus, equations (4.43), (4.45), and (4.46) provide the necessary results if v_j is replaced by u_j and p_i by q_i. If the contact of dissimilar materials is to be addressed the coupling terms (i.e. tangential displacement due to normal load and normal displacement due to tangential load are also needed). For the triangular elements discussed above the required results are

$$\frac{v_j}{q_i} = \frac{u_j}{p_i} = -A\beta \text{sgn}(x_j)\left[c - \frac{(c - x_j\text{sgn}(x_j))^2}{c}\right] + \text{constant} \qquad |x_j| < c \quad (4.47)$$

and

$$\frac{v_j}{q_i} = \frac{u_j}{p_i} = -A\beta c\, \text{sgn}(x_j) + \text{constant} \qquad |x_j| > c \qquad (4.48)$$

When once the contact pressure is known, the interior stress field may be found by superposition, using the solution for the basic pressure element used as a starting point. Details of the interior stress fields needed are given by Hills, Nowell and Sackfield (1993).

The extension of the technique to three-dimensional problems is relatively straightforward in principle although the solution of frictional problems can present some difficulty since shear tractions can now act in any direction. Under these circumstances the variational method of solution might be preferred, despite its mathematical complexity, since stick/slip zones are established automatically. Various traction elements are available for three dimensional problems, but the most commonly used are those of uniform traction acting over a rectangular or square area. The relevant influence functions can be determined form the work of Love (1929) and are summarised by Hills, Nowell and Sackfield (1993). The three-dimensional analogue of the overlapping triangular elements described above is a mesh of hexagonally-based pyramids of traction. Unfortunately, not all of the required displacement components are available in closed form.

4.4.3 Other techniques

All of the procedures described above extend the repertoire of solutions which may be found, but nevertheless retain the assumptions that the contacting bodies are semi-infinite in extent, and that the characteristic radius of curvature of the bodies is very large compared with the size of the contact patch. Whilst these conditions are not always as restrictive as they appear, there are occasions where the influence of far boundaries is significant. For this class of problem there is often little option but to employ finite element or boundary element methods. It is not appropriate to give a comprehensive treatment of these two general methods of stress analysis here but a few comments regarding their use for the analysis of fretting fatigue can be made. In recent years commercial finite element codes have made considerable progress in implementing contact algorithms. For frictionless contact these often work quite well although more difficulties are encountered if shear tractions are to be included. Partial slip problems are particularly taxing as the traction distribution is history dependent and the entire loading history must be followed. This means that solution times can be quite long and moderately powerful computers are required, particularly if residual stress and crack closure effects are to be considered. Finite element methods have been used satisfactorily to analyse fretting problems (e.g. Ruiz and Chen (1986) or Hattori et al.(1988)) although they are not ideal for parametric studies as the domains need to be re-meshed if the crack or contact geometry is altered.

An alternative approach is to employ boundary elements. This method has certain similarities to the influence function technique described above, but is more general in scope. Comprehensive introductions to the method, including its use for contact problems, are given by Crouch and Starfield (1983) or Becker (1992). Several analyses of fretting problems have been conducted (e.g. Sato, (1992)). Reviews of the range of computational techniques available for contact problems, including both boundary and finite element approaches, are given by Kikuchi (1993) and Aliabadi and Brebbia (1993).

4.5 Numerical methods: layered problems

One practical geometry of interest in the analysis of fretting fatigue experiments is that shown in fig. 4.15, where fretting pads contact a plane specimen of finite thickness. At first sight it is not apparent that specimen can be modelled as a half-plane. Clearly, if the specimen thickness is large compared to the contact width, the approximation will be reasonable but will be less good for thin specimens. In designing and analysing experiments it is important to understand the limits of the half-plane approximation and to use a more sophisticated model where required. Fortunately it is relatively straightforward to obtain influence functions for the displacement caused by traction elements acting on an elastic strip of finite thickness and thus the influence function method outlined above can be used to investigate the effect of specimen thickness. Using the origin shown in fig. 4.15, we can obtain the boundary conditions required on $y = 0$ from symmetry as

$$\tau_{xy} = v = 0 \tag{4.49}$$

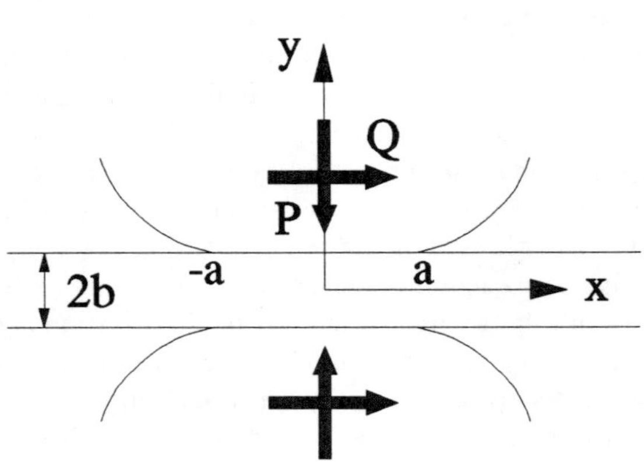

Figure 4.15: Fretting fatigue experimental configuration showing finite specimen thickness.

It will be seen that these are identical to those for the configuration shown in fig. 4.16, where a strip of thickness b rests on a rigid frictionless substrate. The required influence functions for traction elements acting on the surface of the strip ($y = b$) can be obtained using Fourier Transform techniques (see Sneddon, 1951). For a distribution of normal pressure which is even about $x = 0$, the displacement components at the surface ($y = b$) are given by

$$v(x) = \frac{\kappa+1}{2\pi\mu}\int_0^\infty \frac{\sinh^2(\alpha b)}{\alpha b + \frac{1}{2}\sinh(2\alpha b)}\tilde{p}(\alpha)\cos(\alpha x)\frac{d\alpha}{\alpha} \tag{4.50}$$

$$u(x) = \frac{-1}{2\pi\mu}\int_0^\infty \left[(\kappa-1) - (\kappa+1)\frac{\sinh^2(\alpha b)}{\alpha b + \frac{1}{2}\sinh(2\alpha b)}\right]\tilde{p}(\alpha)\sin(\alpha x)\frac{d\alpha}{\alpha} \tag{4.51}$$

where μ is the modulus of rigidity of the strip, $\kappa = 3 - 4\nu$ in plane strain and $(3-\nu)/(1+\nu)$ in plane stress (ν being Poisson's ratio for the strip), and $\tilde{p}(\alpha)$ is the Fourier cosine transform of the applied pressure, defined by

$$\tilde{p}(\alpha) = \int_0^\infty p(x)\cos(\alpha x)\,dx \tag{4.52}$$

Similarly, the displacement components due to an element of shear traction, also symmetrical about $x = 0$ are

$$v(x) = \frac{1}{2\pi\mu}\int_0^\infty \left[(\kappa-1) - (\kappa+1)\frac{\alpha b}{\alpha b + \frac{1}{2}\sinh(2\alpha b)}\right]\tilde{q}(\alpha)\sin(\alpha x)\frac{d\alpha}{\alpha} \tag{4.53}$$

$$u(x) = \frac{\kappa+1}{2\pi\mu}\int_0^\infty \frac{\cosh^2(\alpha b)}{\alpha b + \frac{1}{2}\sinh(2\alpha b)}\tilde{q}(\alpha)\left[\cos(\alpha x) - 1\right]\frac{d\alpha}{\alpha} \tag{4.54}$$

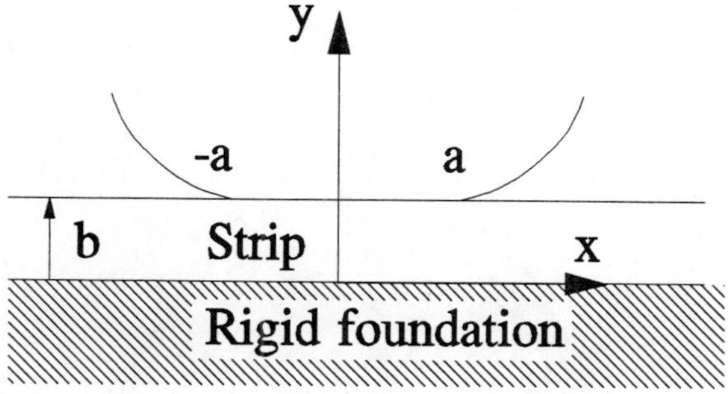

Figure 4.16: An elastic strip of thickness b resting on a frictionless rigid foundation.

where $\tilde{q}(\alpha)$ is the Fourier cosine transform of $q(x)$. For a triangular traction element of peak magnitude t_i acting over the region $|x| < c$ the cosine transform is given by

$$\tilde{t}(\alpha) = -\frac{2t_i}{c\alpha^2}\sin^2\left(\frac{c\alpha}{2}\right) \tag{4.55}$$

Substitution of (4.55) into (4.50), (4.51), (4.53) and (4.54) provides the necessary influence functions to analyse the strip geometry. The functions involve the evaluation of an integral over a semi-infinite range and this can only be achieved numerically. The kernel is oscillatory and decaying in magnitude and care must be taken in the choice of quadrature. Some advice is given by Bentall and Johnson (1968) although in practice little difficulty is encountered using modern adaptive routines. Several configurations have been tackled using influence functions of this type, including tractive rolling (Bentall and Johnson, 1968) and application of normal and tangential forces (Nowell and Hills, 1988a). Modified influence functions derived by the same technique can also be used to treat the problem of contact of a layer *bonded* to a rigid substrate (Bentall and Johnson, 1968, Nowell and Hills, 1988b).

The main feature of the results is that a strip or layer is less compliant than a corresponding elastic half-plane. This means that the contact area is less than that occurring for half-planes and the peak pressure correspondingly higher. The effect can be illustrated by examining the results for the case of frictionless normal indentation. Figure 4.17 shows the distribution of contact pressure obtained for normal indentation of an elastic strip of thickness $2b$ by two cylinders having radius R and similar elastic constants to the strip (fig. 4.15). Figure 4.18 summarises the variation of contact size and peak pressure with strip thickness (normalised with respect to the contact semi-width a_∞ occurring in the equivalent Hertzian contact of half-planes). It can be seen that for $b/a > 2.0$ the solution is little different from than predicted by the half-plane approximation. As the strip thickness is reduced to zero the solution approaches that produced by the contact of two elastic cylinders of radius R under

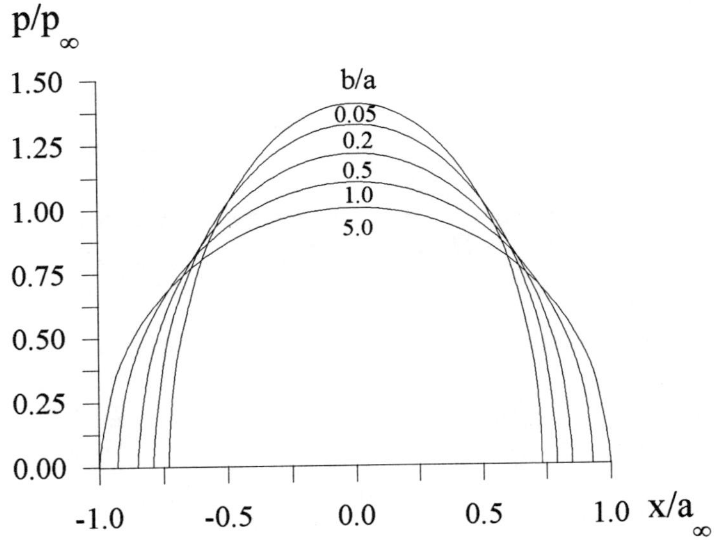

Figure 4.17: Contact pressure distribution for normal indentation of an elastic strip of thickness $2b$ by elastically similar cylinders.

the same normal load, since in this case the strip between the cylinders (fig. 4.15) has negligible compliance. This result is important for the analysis of fretting fatigue since it reveals that the presence of component boundaries has little effect on contact conditions, provided that the boundaries are further than about two or three characteristic contact dimensions away. Hence, many situations can be analysed quite accurately using the half-plane (or in three dimensions the half-space) assumption even if other component boundaries are quite close to the contact.

Stresses within the strip can be conveniently calculated by superposition of the stresses caused by a single traction element. Expressions for the stresses due to a single element of pressure or shear can be obtained using Fourier transform theory (Sneddon, 1951) and are presented in a convenient form by Hills, Nowell and Sackfield (1993).

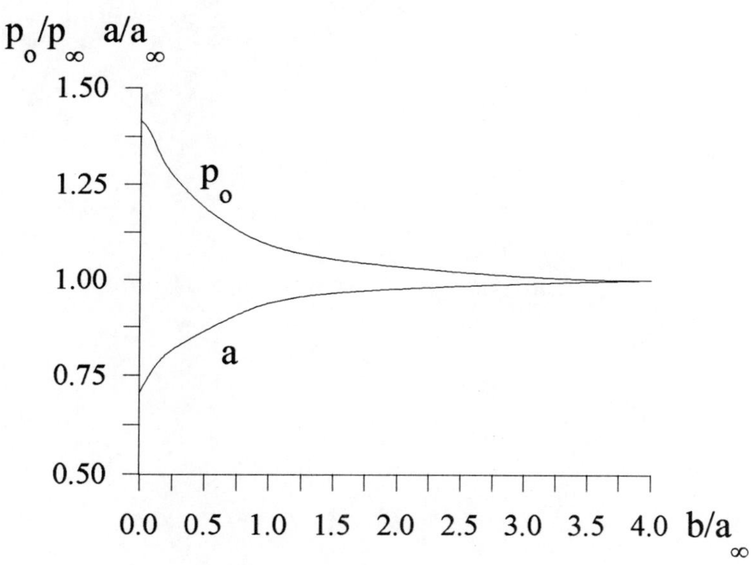

Figure 4.18: Variation of contact size and peak pressure with strip thickness for indentation of a strip of thickness $2b$ by two elastically similar cylinders. Variables are normalised with respect to those occurring in Hertzian contact of a single cylinder with a half plane (a_∞, p_∞).

Chapter 5

Mechanics of Surfaces

5.1 Introduction

The phenomenon of fretting fatigue arises as a result of surface interaction between two contacting bodies and it is therefore important to investigate the nature of the surfaces in some detail in order to understand the phenomenon in any meaningful way. Real surfaces are invariably complex and differ quite markedly from the smooth abrupt transitions from elastic solid to free space which have been assumed so far in this book. For example the surface of a metal component may consist of a number of layers, as depicted in fig. 5.1 for a typical metal. The bulk of the metal component will form a substrate. On top of this will be a layer of material which will typically have been exposed to large amounts of plastic deformation during the manufacturing process. This will have resulted in work hardening and exposure of the material to a temperature cycle so that the yield properties (if not the elastic properties) of this layer may differ significantly from the substrate metal. In addition, the manufacturing process may well have established high residual stresses in this layer and, since all residual stresses need to be self-equilibrating, these effects may persist for some distance from the surface. Most metals will oxidise to some degree and a layer of oxide will normally form on top of the work-hardened layer. The thickness and stability of

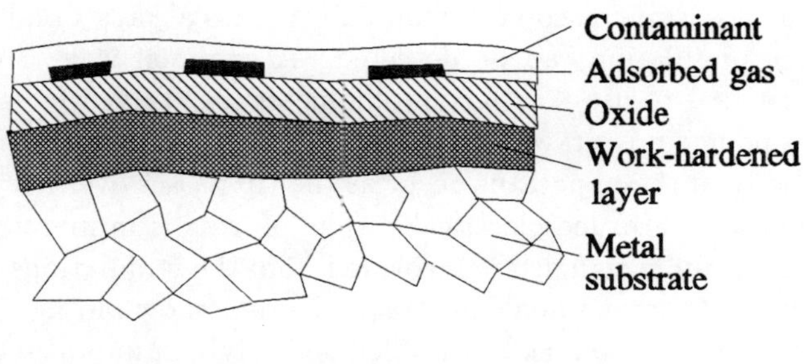

Figure 5.1: Schematic illustration of the surface structure of metals

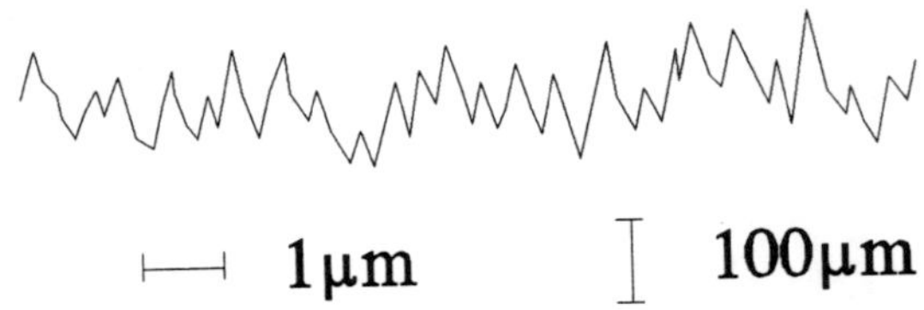

Figure 5.2: Typical profile of a rough surface (note the exaggerated vertical scale).

this layer will depend on the precise chemistry of the metal and of the environment to which it is exposed. Some adsorbed moisture or gas is often present on top of the oxide film and the entire surface may well be covered by a layer of contaminant such as dirt or grease. Surface microstructures of other classes of materials such as polymers or ceramics will differ from those of metals but will be similarly complex.

Faced with such a varied surface microstructure it is perhaps difficult to imagine how any simple model such as an elastic half-plane or half-space can possibly represent a real surface. Fortunately, the contaminant and oxide layers are often quite thin and, whilst they may have a significant effect on the coefficient of friction between two surfaces, their influence on the overall elastic compliance of the contact is small. The work-hardened layer presents more difficulty. However, work-hardening generally affects the yield properties of a material rather than its elastic properties, so that the layer may be taken as having the same elastic constants as the substrate material. Yield properties will, of course, be relevant in cases where surface-initiated material failure is expected, as is the case with fretting fatigue. Residual stresses will also play an important part since any crack growing from the surface will see a combination of applied and residual loading. It is not easy to predict the residual stresses which arise from a particular machining process, but it is possible to obtain estimates of the residual stress distribution from experimental measurements. The topic of residual stresses will be discussed further in §5.4 and also in §6.3 and 6.4, where the incorporation of residual stresses into crack models will be addressed.

In addition to a complex material nature real surfaces also exhibit a complex geometry. No surface is microscopically flat and there is always a scale at which some degree of surface undulation can be detected. A common way of measuring such surface irregularities is to use a surface profilometer (or 'Talysurf') in which a stylus with a small diamond tip is drawn across the surface. Small movements of the stylus relative to the body of the apparatus occur as the tip passes over irregularities in the surface, and these are recorded electrically. The process is in many respects similar to the way in which audio signals are recovered from the undulations of a groove in a gramophone record. Often an analogue trace like that shown in fig. 5.2 is produced, although the signal can, of course, be sampled digitally. An important feature of such traces is the exaggerated vertical scale, which enables the surface irregularities to be

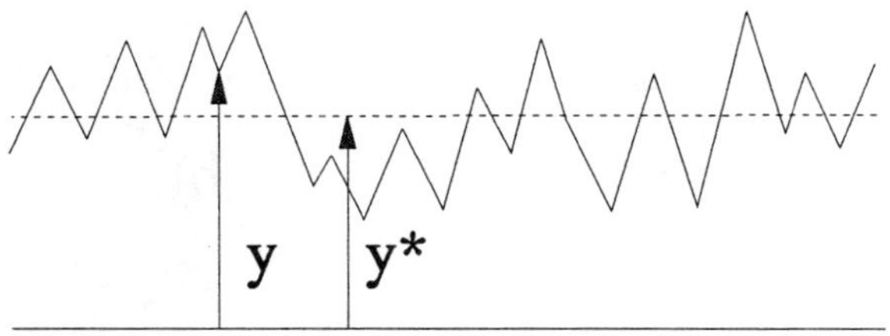

Figure 5.3: Datum, y^*, for the measurement of surface height, y.

clearly seen. However it is important not to gain the impression from this figure that the irregularities themselves are sharp. A more helpful impression of the nature of the surface is perhaps gained by plotting horizontal and vertical motion of the stylus on identical scales.

A number of measures of surface finish are in common use in industry. Two of the most popular are the centre line average roughness, R_a, and the root mean square (RMS) roughness value, R_q. These quantities can be found from a profilometer trace by determining a datum, y^*, such that the areas above and below the datum line are equal (fig. 5.3). This choice is equivalent to minimising the mean square deviation; note that y^* differs from the mean surface height, $\bar{y}$. The centre line average can now be found from the measured profile as

$$R_a = \frac{1}{L}\int_0^L |y - y^*| \, dx \tag{5.1}$$

Similarly the RMS roughness can be evaluated as

$$R_q^2 = \frac{1}{L}\int_0^L (y - y^*)^2 \, dx \tag{5.2}$$

Whilst both these quantities are in common use, they have severe limitations and are most appropriately employed as aids to quality control. As described, the measurements taken are along a single line only. Many surfaces (such as those produced by grinding or turning) have surface features which are strongly orientation-dependent and may not be properly revealed by a trace taken in a single direction. Of course it is possible to build up a more comprehensive picture of a surface by taking several parallel traces and creating a three-dimensional view of surface topography.

A more severe limitation is that R_a, or R_q values give very little information about the nature of the surface. For example the traces depicted in Fig. 5.4 all give rise to identical R_a values yet are profoundly different in character. The measurement technique itself also imposes limitations on the data. The finite size of the stylus means that small features will not be detected (see Hills, Nowell, and Sackfield, 1993) and it is also common practice to eliminate longer wavelength features; if a surface has an overall curvature this will give an artificially high roughness measurement unless

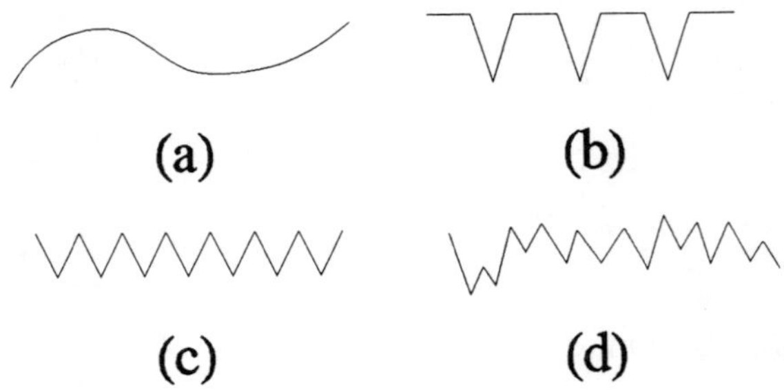

Figure 5.4: Different types of surface finish which give rise to identical R_a values.

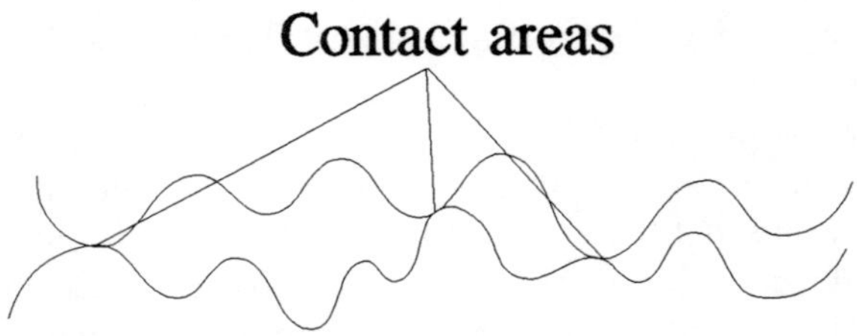

Figure 5.5: Contact of two rough surfaces showing individual asperity contacts.

this wavelength of feature is excluded. Thus, the measurement obtained is a measure of the amount of roughness occurring in a particular bandwidth only.

The phenomenon of surface roughness is clearly important when considering contact between bodies. The most important feature can be illustrated by considering two rough surfaces brought into contact (fig. 5.5). It is apparent that the surfaces contact only at a few points where the peaks, or asperities, of the two surfaces meet. These areas of discrete contact are separated by gaps where no contact occurs and the contact pressure is therefore zero. Contact of rough surfaces can therefore lead to small areas of high contact pressure and larger regions of no contact. All the contact models developed so far have assumed the contacting surfaces to be smooth and the resulting pressure distributions are therefore continuous. The validity of such models for analysing real contacts, where surface roughness is always present to some degree, must therefore be called into question. Fortunately the scale of the surface roughness is often very different from the scale of the overall contact. Thus, although contact occurs only in discrete regions, the result is a global deformation and an average contact pressure which is little different from that predicted by analysis of the corresponding smooth contact. It is therefore possible to invoke St Venant's principle (see Barber, 1992) to deduce that, except for points very close to the surface, the interior stress field within the contacting bodies is almost identical to that predicted by the smooth contact model.

In investigating fretting fatigue it is necessary to consider both the initiation and propagation of cracks under the contact stress field. Once the crack has grown sufficiently so that the tip is some distance from the surface, it will experience loading which is little different from that which would occur in the contact of ideal smooth surfaces, and a smooth contact model will be appropriate. On the other hand, initiation is clearly dependent on surface and near-surface conditions so that the local stress field caused by surface roughness may well be important. In analysing fretting fatigue it is therefore important to consider the scale of surface roughness present in the contact and to evaluate the extent to which it needs to be taken into account in developing a model of the fretting contact. These issues will be addressed in §5.2.

So far the discussion has focused on the effect of surface roughness on the transfer of normal load between the contacting components. Transfer of tangential load is also important in the fretting process and it is usual to develop a friction model to describe the ability of surfaces to effect this load transfer. The commonly used friction models can all be related, to some degree, to surface topography and such relationships will be examined in §5.3. Appropriate friction models for the analysis of fretting problems will be presented, and the determination of friction coefficient in practical configurations addressed.

As stated above, the existence of residual stresses is another important feature of the fretting problem which is related to the surface treatment. In order to incorporate residual stresses correctly into a fretting model it is important to understand several features of residual stress fields. Hence, this chapter will be concluded by a short introduction to the analysis of residual stresses.

5.2 Contact of rough surfaces

There are three possible ways of extending the contact mechanics techniques developed so far to account for surface roughness:

(i) By incorporating a characteristic description of the surface roughness into the contact model. For example, many surfaces, such as those generated by turning or grinding, exhibit periodic undulations which may be incorporated into the analysis by adding a periodic variation of surface height to a smooth model of the overall surface profile.

(ii) By modelling the surface roughness as a random variation in surface height and by investigating the effect of such a variation on the mean pressure developed between two contacting bodies. Such analyses invariably involve the use of statistical methods.

(iii) By taking precise measurements of the surface profiles of actual components and by employing numerical methods, such as those presented in Chapter 4, to predict the contact pressures when the two surfaces are brought into contact in a particular alignment. It is not often that the luxury of such measurements

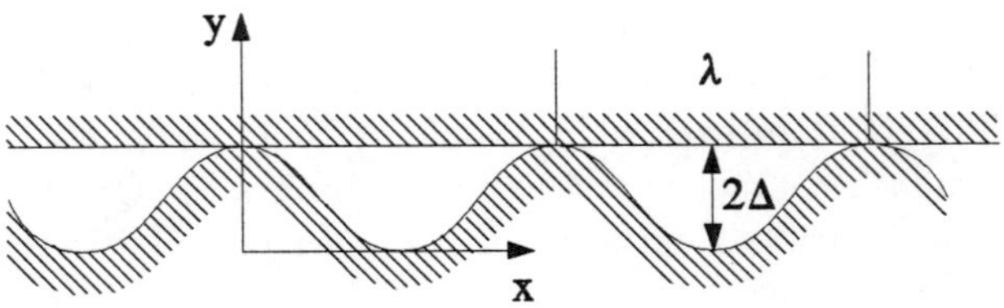

Figure 5.6: Contact of a sinusoidally rough half plane with a smooth surface.

can be justified. Nevertheless the technique does shed some light on the general features of rough contacts and is thus worthy of some discussion.

5.2.1 Regular roughness

We shall commence by considering method (i). Many surfaces exhibit some periodic roughness which can be incorporated into a contact model in a useful way. Consider, for example, the situation shown in fig. 5.6 where a surface which is sinusoidally rough is brought into contact with a rigid flat surface. If the normal load is increased from zero, contact will take place initially only at the peaks of the rough surface. If the contact areas at these peaks are small in relation to the wavelength of the rough surface there will be little interaction between the individual contacts and each individual point of contact can be treated in isolation and is therefore essentially Hertzian. As the load is increased the width of each individual contact becomes significant compared with the wavelength of the undulations and interaction between adjacent contacts must be considered. Eventually, however, if the load is increased still further, the deformation will increase so that the remaining gaps between the surfaces are eliminated and contact takes place over the complete surface. This situation is relatively straightforward to analyse: Consider the configuration given in fig. 5.6 where the surface height, y, can be expressed as

$$y = \Delta(1 + \cos(2\pi x/\lambda)) \tag{5.3}$$

If a pressure, p, of the form

$$p(x) = -p_0 \cos(2\pi x/\lambda) \tag{5.4}$$

is applied to the surface between $x = \pm\infty$, then the resulting displacements can be calculated from equation 2.15 as

$$\frac{\partial v}{\partial x} = \frac{\kappa + 1}{4\pi\mu} \int_{-\infty}^{\infty} \frac{-p_0 \cos(2\pi\xi/\lambda)}{x - \xi} d\xi \tag{5.5}$$

Making the substitution $s = x - \xi$ and carrying out the integration yields

$$\frac{\partial v}{\partial x} = \frac{\kappa + 1}{4\mu} p_0 \sin(2\pi x/\lambda) \tag{5.6}$$

and thus

$$v = \frac{-p_0 \lambda (\kappa + 1)}{8\pi\mu} \cos(2\pi x/\lambda) + C \tag{5.7}$$

where C is an arbitrary constant. Hence, the application of a sinusoidally varying pressure leads to a sinusoidally varying displacement. The gap between the unloaded surfaces in fig. 5.6 is

$$h(x) = \Delta [1 - \cos(2\pi x/\lambda)] \tag{5.8}$$

so that a pressure distribution of the form

$$p(x) = -[\bar{p} + p_0 \cos(2\pi x/\lambda)] \tag{5.9}$$

is required. Here the magnitude of the oscillating component, p_0, is given by

$$p_0 = \frac{8\pi\mu\Delta}{\lambda(\kappa + 1)} \tag{5.10}$$

The mean pressure is clearly $\bar{p}$ and this must be greater than p_0 if tensile tractions are not permitted. For intermediate cases, where $\bar{p}$ is less than this critical value, so that gaps appear between the surfaces, but where interaction between adjacent contacts is important, it is necessary to develop a full solution. Such a solution has been given by Westergaard (1939) who showed that a pressure of the form

$$p(x) = \frac{-2\bar{p}\cos(\pi x/\lambda)}{\sin^2(\pi a/\lambda)} \sqrt{\sin^2(\pi a/\lambda) - \sin^2(\pi x/\lambda)} \tag{5.11}$$

applied to a half-space over contact regions $|x - n\lambda| < a$ together with zero pressure outside these regions gives rise to surface displacements of the form

$$v = \frac{-\bar{p}\lambda(\kappa + 1)}{8\pi\mu \sin^2(\pi a/\lambda)} \cos(2\pi x/\lambda) + C \qquad |x - n\lambda| < a \tag{5.12}$$

within the contact regions. This displacement cancels out the initial gap (5.8) in these regions provided that

$$\bar{p} = \frac{8\pi\mu\Delta \sin^2(\pi a/\lambda)}{\lambda(\kappa + 1)} \tag{5.13}$$

Outside the contact regions the surface displacements are such that the gap remains positive and thus a valid solution exists for contact over regions of width $2a$ at each asperity peak. Figure 5.7 shows the pressure distribution given by equation 5.11 for various different widths of contact. It can readily be seen that the pressure distribution becomes the sinusoidal variation predicted by equation 5.9 when $a = \lambda/2$ and the gaps are eliminated whereas as a/λ becomes small the elliptical Hertzian pressure distribution is recovered.

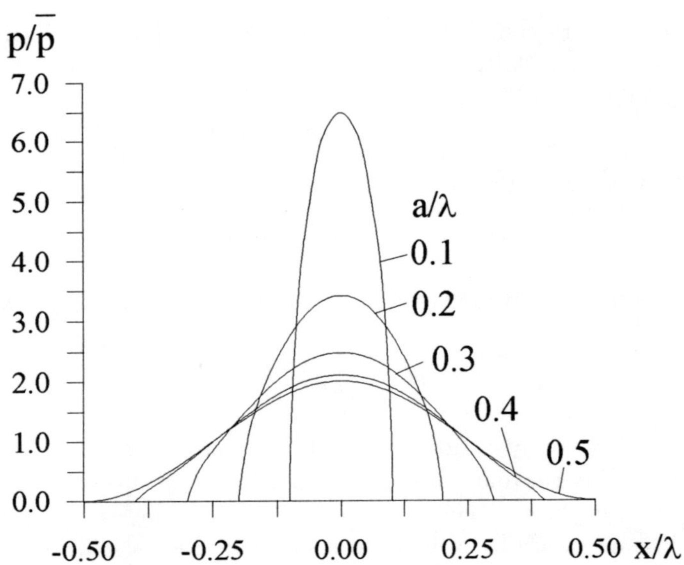

Figure 5.7: Pressure distributions found by Westergaard for contact between sinusoidally rough surfaces.

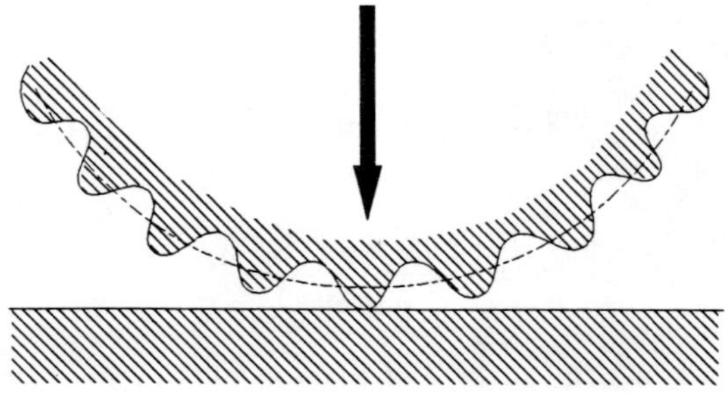

Figure 5.8: Sinusoidal roughness existing on a cylindrical body in contact with a flat surface.

Consideration so far has focused on surfaces which exhibit surface roughness superimposed on a flat plane. Of perhaps more practical interest for the fretting problem are cases where a degree of roughness exists together with some overall variation in surface height. Consider the geometry shown in fig. 5.8 where a sinusoidal surface roughness is present on a cylindrical body. This might be taken to represent, for example, a cylindrical fretting pad produced by grinding. If the roughness is small, there is likely to be complete contact so that there are no gaps between the cylinder and a flat surface when the two surfaces are brought together under a normal load. In this case the configuration may be simply analysed by adding the surface roughness to the underlying smooth cylindrical surface profile. The pressure distribution may then be obtained in the usual manner from integral equation 2.22, i.e.

$$\frac{1}{A}\frac{\partial h}{\partial x} = \frac{1}{\pi}\int_{-a}^{a}\frac{p(\xi)}{x-\xi}\,d\xi \qquad |x| \le a \tag{5.14}$$

In this case the gap function, h, may be expressed as

$$h(x) = -\frac{1}{2}kx^2 - \Delta\left[1-\cos\left(2\pi x/\lambda\right)\right] \tag{5.15}$$

where k is the curvature of the underlying surface profile. Substitution into equation (5.14) yields

$$\frac{1}{\pi}\int_{-a}^{a}\frac{p(\xi)}{x-\xi}\,d\xi = -\frac{kx}{A} - \frac{2\pi\Delta}{A\lambda}\sin\left(2\pi x/\lambda\right) \qquad |x| \le a \tag{5.16}$$

This equation cannot readily be inverted analytically, but the numerical techniques discussed in §4.4.1 enable a solution to be found quite simply. A typical pressure distribution is given in fig. 5.9, where it can be seen that the result is quite similar to that obtained for a flat surface, namely that the overall result for a smooth geometry (in this case the elliptical Hertzian distribution) has a sinusoidal component of pressure superimposed. If the amplitude of the surface roughness is increased (or the wavelength reduced) the size of the sinusoidal component of pressure increases and tensile surface tractions are predicted at some positions. For this case a more sophisticated analysis incorporating multiple contact patches is required. Such a solution would not be easy to obtain. However it is possible to produce a solution for the case where the areas of contact are widely spaced. Under these circumstances each asperity contact may be treated as Hertzian since interaction between the individual contacts will be limited. An analysis along these lines was first presented by Nowell and Hills (1989). Consider the situation of fig. 5.8 once more and suppose that the centre of the overall contact coincides with an asperity tip so that the situation is symmetric and a total of $2N+1$ asperities are in contact, each of radius of curvature ρ. The surfaces must touch at the centre of each asperity and this means that the bulk deformation necessary at these points is the same as that required for the equivalent

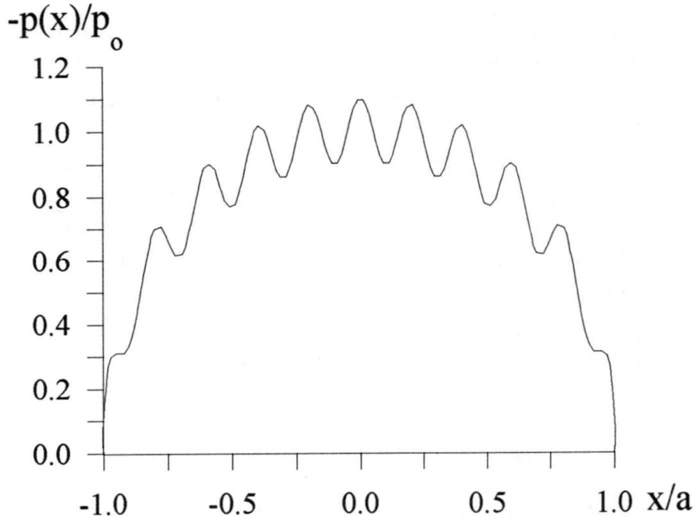

Figure 5.9: Typical pressure distribution resulting from contact of a cylinder having sinusoidal roughness with a flat surface for the case where continuous contact occurs.

smooth contact so that

$$h(\lambda n) = -\frac{1}{2}k(\lambda n)^2 \qquad n = -N, ...N \tag{5.17}$$

Suppose that the nth contact carries a load P_n distributed as a Hertzian pressure over the region $|x - \lambda n| \leq a_n$. The displacements due to this load can be obtained from equations 2.57 and 2.58 as

$$v_n(x) = -\frac{AP_n}{\pi}\left[\left(\frac{x-\lambda n}{a_n}\right)^2 + C_0\right] \qquad |x - \lambda n| \leq a_n \tag{5.18}$$

inside the nth contact and

$$v_n(x) = -\frac{AP_n}{\pi}\left[\ln|X| + \frac{1}{2X^2} + C_1\right] \qquad |x - \lambda n| > a_n \tag{5.19}$$

outside, where

$$X = \left|\frac{x-\lambda n}{a_n}\right| + \sqrt{\left(\frac{x-\lambda n}{a_n}\right)^2 - 1} \tag{5.20}$$

Continuity of displacement at $x = \lambda n \pm a_n$ is ensured by writing $C_0 = C_1 - \frac{1}{2}$; the remaining constant is removed by referring displacements relative to the origin, $x = 0$, i.e. defining

$$v_n^*(x) = v_n(x) - v_n(0) \tag{5.21}$$

The total relative displacement due to all contacting asperities may now be found by summing the individual contributions so that

$$v^*(x) = \sum_{n=-N}^{N} v_n^*(x) \tag{5.22}$$

The load on each individual contact is determined by enforcing the required relative displacement at the centre of each asperity (equation 5.17) so that

$$v^*(\lambda n) = \frac{-k(\lambda n)^2}{2} \qquad n = -N, ..., -1, 1, ...N \tag{5.23}$$

Note that there is no equation corresponding to $n = 0$ because relative displacements have been used and thus displacements are automatically matched at the origin. The remaining equation required for solution is obtained from overall equilibrium with the applied load P so that

$$\sum_{n=-N}^{N} P_n = P \tag{5.24}$$

The equations may be solved by sequential iteration, noting that P_n and a_n are related by the Hertzian contact law (equation 2.36) so that

$$a_n^2 = \frac{2P_n \rho a}{\pi} \tag{5.25}$$

Initial estimates are chosen for N, P_n, and a_n and equations (5.23) and (5.24) used to obtain new estimates for P_n and a_n. The iteration continues until convergence occurs. It is then possible to check that the correct number of contacts has been specified for the overall load by ensuring that no asperity loads are tensile and that no interpenetration of material takes place outside the Nth contact. Note that the solution technique has only matched displacements explicitly at the centre of each asperity, but if the asperities are sufficiently widely spaced, the displacements at one particular asperity due to the influence of the *other* contacts will not vary greatly over the width of that asperity contact patch. The local variation in displacement required at each individual contact is automatically provided by assuming Hertzian contact. A typical pressure distribution obtained from this analysis is shown in fig. 5.10. It can clearly be seen that the central asperities carry higher loads, as might be expected from the smooth Hertzian analysis. Similar configurations have been examined experimentally (using photoelasticity) by Liebensperger and Brittain (1973) and numerically using the finite element method by Komvopoulos and Choi (1992). In each case broadly similar results were obtained.

5.2.2 Random rough surfaces

The analysis discussed so far is appropriate where the roughness of the contacting bodies is regular in some characteristic way. Not all rough bodies exhibit such roughness; many rough surfaces, such as those produced by shot peening or abrading have a

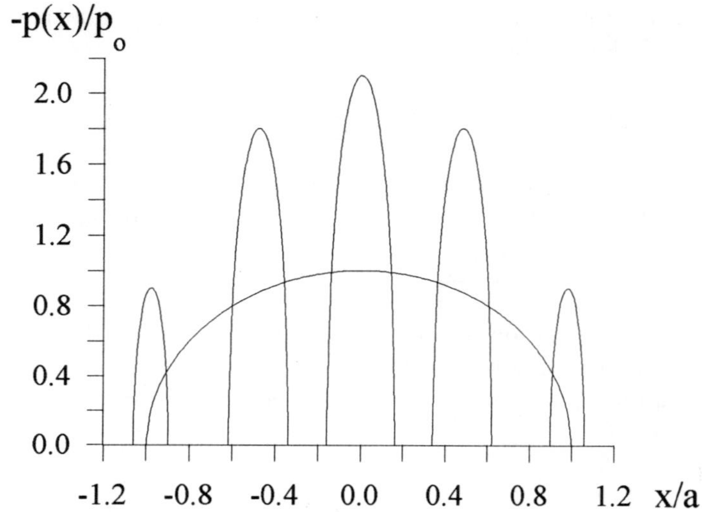

Figure 5.10: Typical pressure distribution resulting from contact of a regularly rough cylinder with a smooth surface for the case where discontinuous contact occurs.

distribution of surface heights which is effectively random. Under these circumstances two avenues of approach are available: either the profiles of the two particular contacting bodies under consideration must be measured and the contact pressure when they are brought into contact in a particular orientation determined numerically, or a statistical analysis may be carried out in order to highlight some general features of the contact of random rough surfaces which may be expected to arise. The former approach does not shed much light on the overall attributes of this type of contact and it is therefore appropriate to begin by examining the statistical approach. The analysis carried out in the previous section has shown that in the presence of a moderate amount of surface roughness the contact is divided into zones of contact, with high contact pressures, separated by areas where no contact occurs and the pressure is zero. A statistical approach is not capable, of course, of determining the actual contact pressure between a pair of contacting bodies, but instead reveals the mean, or statistically expected, pressure as a function of position in the contact if a large number of geometrically similar bodies[1] are brought into contact under identical normal loads. If the number of asperity contacts in a particular overall contact is large, then a moving average of the actual contact pressure would smooth out the local variations, and the resultant average pressure distribution will approach the statistically expected value. This statistically expected pressure is often termed the *effective pressure* and it is important to understand the distinction between it and the actual contact pressures which occur between a particular pair of rough bodies. A calcula-

[1]In a macroscopic sense; the roughness will vary in a random manner.

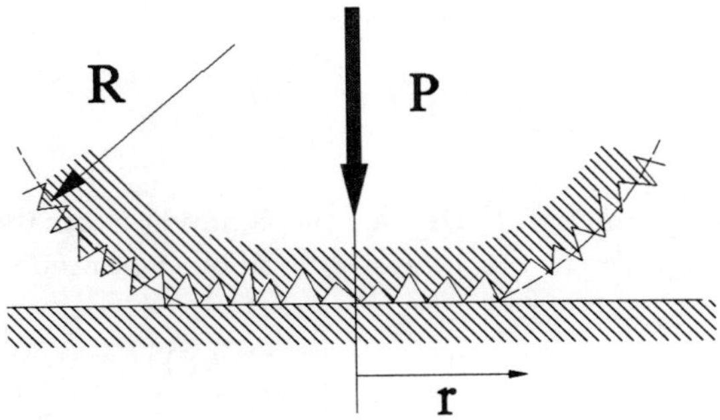

Figure 5.11: A rough elastic sphere in contact with a smooth rigid half-space.

tion of the effective pressure is useful because St Venant's principle will mean that the influence of individual asperity contacts on subsurface stresses will be confined to a small surface layer and hence the bulk stress field resulting from the overall contact will depend on the expected pressure distribution alone.

It might be thought at first that the effective pressure distribution will differ little from that obtained for an equivalent smooth contact. However two considerations show that this is not necessarily so. First the rough surface layer will be more compliant than if it were composed from solid material, and it is clear that material compliance has a significant effect on contact pressure. Secondly, suppose that the asperity heights follow some statistical distribution such as the Normal distribution function. There is always a small but finite possibility that an asperity height several standard deviations from the mean will occur. When a rough body with such a random distribution of asperity heights is brought into contact with another body there is therefore a small but finite probability that the gap between the bodies at some large distance from the centre of the contact will be bridged by a high asperity. The statistically expected pressure cannot be zero at this point and it might therefore be expected that the pressure distribution would be spread over a larger area than in the case of smooth contact.

The first successful analysis of the contact of random rough bodies was carried out by Greenwood and Tripp (1967). They examined the case shown in fig. 5.11, where a rough elastic sphere is brought into contact with a smooth rigid half space. The confining of both elasticity and roughness to one of the two contacting bodies simplifies the analysis and involves no loss of generality since it is always possible to model the more general case of two rough elastic bodies by adjusting the elastic compliance and roughness of the model. The asperity peaks on the sphere are assumed to have a *constant* radius of curvature, ρ, and are small compared with the radius of the sphere, R, so that many asperities are in contact. The peaks of the asperities are assumed to have heights which are distributed randomly according to a probability density function ϕ, so that the probability of the height, ζ_i, of a particular asperity lying in the interval $\zeta_1 < \zeta_i < \zeta_2$ is given by

$$\Pr(\zeta_1 < \zeta_i < \zeta_2) = \int_{\zeta_1}^{\zeta_2} \phi(\zeta)\, d\zeta \tag{5.26}$$

The probability density function (PDF) for the common Normal or Gaussian distribution, which is appropriate to the analysis of a number of random rough surfaces is given by

$$\phi(\zeta) = \frac{1}{\sigma\sqrt{2\pi}} \exp\left[\frac{-\left(\zeta - \bar{\zeta}\right)^2}{2\sigma^2}\right] \tag{5.27}$$

where $\bar{\zeta}$ and σ are the mean and standard deviation respectively of the distribution. The difference in order of size between the asperities and the bulk contact means that the elastic deformation occurring at a particular asperity may be thought of as resulting from two sources:

1. Overall or bulk deformation, v^b, of the sphere as an elastic half-space under the action of the expected pressure.

2. Local deformation, v^a, in the individual asperities under the asperity load.

Consider the centre of the sphere displaced towards the plane by a distance Δ. The required displacement at an asperity peak of height ζ_i situated at a radial distance r from the centre of the contact is given by

$$v(r) = \Delta - \frac{r^2}{2R} + \zeta_i \tag{5.28}$$

where the usual Hertzian approximation of representing the spherical surface as a parabaloid has been employed. This displacement must be accommodated by the sum of bulk and asperity displacement components so that

$$v_i^a + v^b(r) = \Delta - \frac{r^2}{2R} + \zeta_i \tag{5.29}$$

and thus the asperity component is

$$v_i^a = \Delta - \frac{r^2}{2R} + \zeta_i - v^b(r) \tag{5.30}$$

If the individual asperity contacts are modelled as Hertzian, the asperity deformation may be related to the load using equation 2.116 with one body taken as rigid to give

$$P_i = \frac{8\mu}{3(1-\nu)} \rho^{\frac{1}{2}} \left(v_i^a\right)^{\frac{3}{2}} \tag{5.31}$$

It is now possible to calculate the effective pressure, p^*, at a particular radius r by integrating the product of the load for a particular asperity height and the probability of its occurrence over the range of possible contacting asperities to give

$$p^*(r) = \frac{8\mu}{3(1-\nu)}\rho^{\frac{1}{2}} N \int_{\delta}^{\infty} (v^a(\zeta))^{\frac{3}{2}}\, \phi(\zeta)\, d\zeta \qquad (5.32)$$

where N is the number of asperities per unit contact area and the lower limit of integration, δ, corresponding to the asperity height which just bridges the gap between the sphere and the plane, is given by

$$\delta = v^b(r) + \frac{r^2}{2R} - \Delta \qquad (5.33)$$

Greenwood and Tripp carried out the integration in equation (5.32) numerically for a Gaussian distribution of peak heights. They found that the pressure distribution obtained depended on two dimensionless variables:

$$T = \frac{P(1-\nu)}{\mu R_a \sqrt{2RR_a}} \qquad (5.34)$$

and

$$M = \frac{8}{3} N R_a \sqrt{2R\rho} \qquad (5.35)$$

Where R_a is the centre line average roughness (5.1). Johnson (1985) discusses these results and shows that T is a measure of the ratio of roughness present to the compliance of an elastically similar smooth sphere under the same normal load. The second parameter, M, is independent of load and is a measure of the topography of the rough surface. Figure 5.12 shows typical distributions of effective pressure. At low values of T, corresponding to low loads or high roughness, the distribution of effective pressure exhibits a characteristic bell shape and the maximum effective pressure is much lower than the Hertzian value. At higher values of T, the distribution of effective pressure approaches the smooth Hertzian result although it must be emphasised that the *actual* pressures still exhibits high peaks at the asperities separated by areas of no contact; it is merely the mean or expected value which approaches the Hertzian case. The maximum effective pressure is found to be a strong function of T but a relatively weak function of M.

Discussion of the overall area in which asperity contact occurs is more complicated since, as noted above, there is always a finite probability of contact even at large distances of the centre of contact. The effective pressure distribution never, therefore, falls to zero and an arbitrary definition of the edge of contact must be made. Greenwood and Tripp define an effective contact radius a^* given by

$$a^* = \frac{3\pi \int_0^{\infty} r p^*(r)\, dr}{4 \int_0^{\infty} p^*(r)\, dr} \qquad (5.36)$$

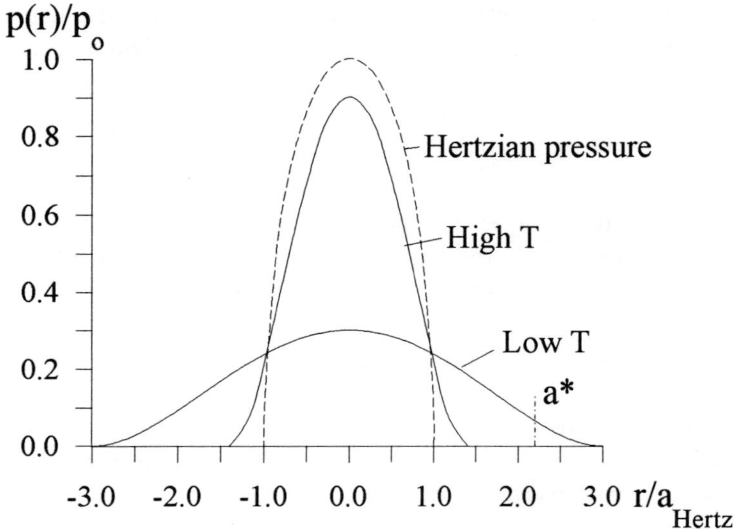

Figure 5.12: Typical effective pressure distributions for contact of a randomly rough sphere with a rigid plane surface.

and this has the virtue that a^* approaches the Hertzian contact radius as the roughness tends to zero ($T \rightarrow \infty$). Using this definition, the effective radius of rough contact is always greater than for the corresponding smooth configuration. Greenwood and Tripp's analysis addresses the spherical configuration, but a similar analysis has been carried out for the contact of rough cylinders by Lo (1969). Lo's model assumes axisymmetric asperity peaks with a constant radius of curvature. Drawing on the results of Greenwood and Tripp, he assumes a bell-shaped distribution of effective pressure having the form

$$p^*(x) = -p_0^* \left[1 - (x/a^*)^2\right]^2 \tag{5.37}$$

This simplifies the analysis but it means that the solution obtained does not approach the equivalent smooth case as the roughness is reduced.

Two fundamental effects of surface roughness have now been demonstrated: the actual pressure distribution is made up of regions of high pressure at the asperity tips together with regions of no contact in between; the distribution of mean or effective pressure may also be modified by the presence of surface roughness. We now turn to the third analysis technique, that of the analysis of real rough surfaces.

5.2.3 Real rough surfaces

The third useful approach to the analysis of rough contacts is to take measurements of two specific rough surfaces and to analyse the situation when they are brought into contact in a specific orientation. It is not often that such detailed measurements can be justified and even then, a single analysis only applies to a particular orientation; if the alignment is slightly disturbed a completely different distribution of asperity contact pressures may occur. Thus, such analyses are not particularly useful in analysing a real contacts since a small change in operating conditions will alter the result significantly. Nevertheless they can be useful in that they demonstrate in a graphic way the form of typical contact pressure distributions between rough surfaces. Many papers have been published on the subject, but of particular note are those by Webster and Sayles (1986) and West and Sayles (1988). These treat two and three dimensional configurations respectively. The technique employed is to divide the surface into a number of regions over which individual pressure elements (of the form discussed in §4.4) are chosen to act. The pressures are then adjusted by an iterative method to ensure that there is no interpenetration of material and that no tensile tractions arise. The resulting pressure distributions are very similar to those obtained for regular rough surfaces in that there are sharp peaks of pressure separated by areas where there is no contact.

5.2.4 Stresses under rough contacts

It will be apparent from the foregoing discussion that the surface tractions occurring under rough contacts can be very different from those obtained from a smooth contact model, both in the sense that the actual contact load is concentrated into localised asperity contacts and also in that the mean or effective pressure obtained when the asperity loads are averaged out may also be affected. Since all real surfaces are rough to some degree it is necessary to examine whether a smooth contact model can ever be a valid representation of contact conditions. Fortunately the scale of the overall contact is often very different from that of the individual asperity contacts. In analysing fretting fatigue the subsurface stresses are often of primary concern, particularly during the propagation phase of component life[2]. From St Venant's principle it might be expected that the stresses resulting from individual asperity contacts will only persist for distances of the same order as the scale of asperity contact. At larger distances from the surface, the overall bulk load will determine the stress field. In effect there are two distinct regions in the subsurface stress field. There is a surface layer with a thickness of the order of the asperity contact dimension in which stresses caused by the individual asperity contacts dominate. At larger depths the stresses may be considered as arising from the overall contact rather than from the contact of many individual asperities and are thus similar to those which would be generated by the effective pressure distribution. This phenomenon is best illustrated by calculating some typical subsurface stress fields for rough contacts. Consider the analysis presented in §5.2.1. Here the contact pressure consists of a number of

[2] As opposed to the initiation phase, where we expect that conditions on the surface will control component behaviour.

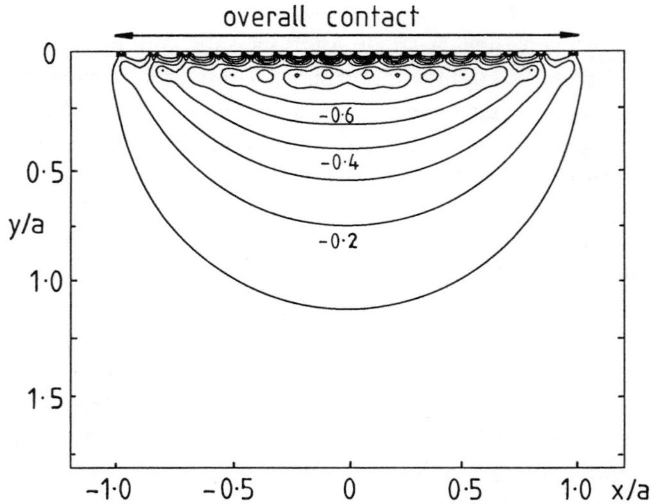

Figure 5.13: Contours of most positive principal stress under a typical rough contact. Stresses are normalised with respect to the maximim pressure in an equivalent smooth contact.

individual Hertzian distributions. The stress field beneath the overall contact can be calculated by superposing the results for the individual asperity contacts using the equations for the stresses beneath a Hertzian distribution of normal pressure (2.44). A typical plot is shown in fig. 5.13 for a contact where there are 15 individual asperity contacts. Local maxima are clearly visible under the individual asperities but the global stress field is very similar to that for the equivalent smooth contact (fig. 5.14) Very similar results have been found by Komvopoulos and Choi (1992), using the finite element method. In practice the number of asperity contacts is likely to be much higher than 15 and the layer of perturbed stresses correspondingly thinner.

It is clear therefore that smooth contact models are useful in predicting subsurface stresses away from a thin surface layer. In analysing fretting fatigue it is necessary to consider whether a smooth or rough surface model is more appropriate. If the analysis of relatively long fatigue cracks is contemplated (see Chapter 6) then the crack tip will be well away from the surface layer where asperity stresses dominate and a good representation of the stress intensity factor will be obtained by using the stresses resulting from a smooth contact model. For short cracks, where the crack length is of the same order as the scale of the asperities such an approximation will be poor and it will be important to include some asperity effects in the crack model. Miller, Keer, and Cheng (1985) have presented an analysis of cracks under contact loading in which asperity and bulk effects are both included and the relative contributions of each can be assessed. Surface effects will be dominant in the initiation phase of crack life and it is important to consider the effects of asperities during this part of a component's

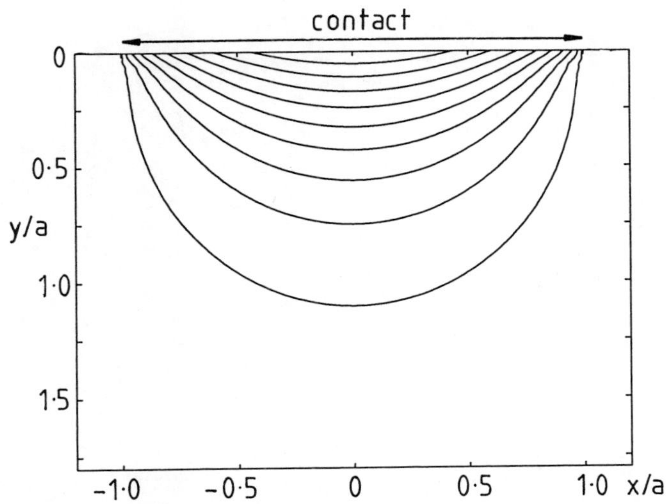

Figure 5.14: Contours of most positive principal stress for the smooth contact equivalent of figure 5.13.

loading history. Appropriate models for crack initiation will be discussed further in Chapter 9.

5.3 Friction

Frictional or resistive shearing forces arise whenever two bodies are pressed into contact and one slides over the other. The direction of the friction force always precisely opposes the relative motion. The notion of friction was introduced almost 300 years ago by Amontons, who also postulated two useful 'laws' which are borne out by observation of the contact of rigid bodies. These are:

1. The friction force is independent of the apparent surface area in contact. Thus, if a heavy block with unequal length edges is pushed along a table the resistive force developed is the same, regardless of which face is in contact.

2. The friction force is proportional to the normal load, i.e.,

$$T = fN \tag{5.38}$$

   where N is the normal load, T the force resisting motion and f is the coefficient of friction.

Sometimes the coefficient of friction is taken to be a constant for a particular pair of materials, but this is an idealisation, and a more representative behaviour is shown

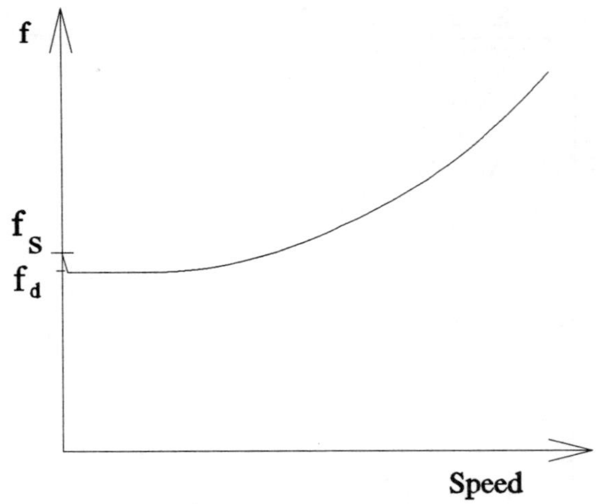

Figure 5.15: Variation of coefficient of friction with sliding speed. Note the difference between static, f_s, and dynamic, f_d, values.

in fig. 5.15. This shows that for a pair of bodies in contact and at rest, the maximum frictional force developed will be given by

$$T = f_s N \tag{5.39}$$

where f_s is the coefficient of static friction. As soon as a tangential force equal to the limiting value is applied, the bodies will start to slide, and when this occurs the frictional force will generally fall to a new value

$$T = f_d N \tag{5.40}$$

where the coefficient of dynamic friction f_d is less than f_s. If the speed of sliding is increased significantly for many material pairs, the frictional force will also increase; however, for a wide range of relative velocities the frictional force remains constant, and where this is so Coulomb's Law of friction is said to hold: the friction force is independent of relative sliding velocity.

Our present understanding of the origin of frictional forces does not permit a systematic prediction of f for a given pair of materials. It is, instead, safest to measure the frictional force developed between the two bodies being slid together experimentally, ideally under conditions of closely controlled imposed relative velocity, to avoid stick-slip phenomena associated with the difference between f_s and f_d. Further, the physical condition of the surfaces being slid together has a profound influence on the coefficient of friction. Specifically, the friction depends on;

1. Surface Topography. Friction has its origin both in the mechanical interlocking of surface asperities and in the form of adhesive bonding which occurs at areas of very high local contact pressure. A change in surface roughness may therefore be expected to modify the friction force. Furthermore, those surfaces which have a significant directionality to their texture, perhaps produced by surface grinding, may be expected to have rather different coefficients of friction in different directions.

2. Cleanliness of surfaces. All real surfaces are lubricated by contaminants, but after extremely careful cleansing and degreasing it is possible to achieve high coefficients of friction between components.

The above observations suggest that frictional effects are variable from point to point within the contact, and are capricious in nature. It is therefore quite surprising that the Laws of Amontons and Coulomb hold, although it ought to be emphasized that they are consistent *only for a given pair* of bodies. It is possible to explain Amontons' Laws in terms of known properties of surface topography. First, if the *real* area of contact[3] between a pair of rough sliding bodies is A_r, the frictional shearing traction, τ_f, is given by

$$\tau_f = \frac{T}{A_r} \tag{5.41}$$

Secondly, Greenwood and Williamson (1966) have shown that under a wide range of conditions the real area of contact is proportional to the normal load, i.e.,

$$A_r = kN \tag{5.42}$$

By eliminating the real area of contact, A_r between these equations we see that

$$T = k\tau_f N \tag{5.43}$$

Equation (5.43) indicates that the friction force is proportional to the normal load and is independent of the apparent area of contact, providing that τ_f is a property of the surface-pair. Possible explanations of the genesis of friction are summarized by Arnell et al. (1991).

Equations (5.41) and (5.42) provide an insight into possible extensions of the definitions of friction, so that they may profitably be applied not globally to the contact as a whole, but to regions within the contact, as described in §3.1. It is reasonable to assume that when two bodies are first brought into contact, providing the surfaces were cleaned and machined uniformly, the ability of any localized area to resist relative motion is independent of position within the contact. However, if the two bodies then suffer fretting, the contact zone will consist of zones of adhesion (in the sense that no relative motion will occur) where little surface modification

[3]That is, the sum of the individual asperity contact areas.

is anticipated, and zone of relative slip, where plastic flow and a modification of surface topography is inevitable. Normally, the modification which occurs is such as to *increase* the coefficient of friction, so that some regions zones of microslip may, in time, give way to stick. In problems of this kind, it is important to emphasise the following points:

1. Within zones of stick the coefficient of friction is statically indeterminate. If the direct and shear components of traction sustained at some point are p and q respectively then they may be related by

$$q = \lambda(x)p \tag{5.44}$$

and the quantity $\lambda \leq f$. The equality can only occur when particles are on the verge of slipping, and in turn this can obtain only on the boundary between slip and stick regions.

2. Within slip zones a generalization of equations (5.38) suggests itself, viz.

$$q(x) = f(x)p(x) \tag{5.45}$$

Note that there is no requirement in this equation for f to be constant over the slip zone, either in time or spatially, and that this pointwise definition of f is consistent with a macroscopic definition (equation (5.38)).

3. If the coefficient of friction defined by (5.45) is adopted, the macroscopic equivalent may be thought of as some average or mean coefficient, $\bar{f}$, which might then be defined by

$$\bar{f} = \frac{\int q\,dA}{\int p\,dA} \tag{5.46}$$

where the integrals are taken over the area of contact. It follows that because f is indeterminate within stick zones and $\bar{f}$ is the only macroscopically observable or measured quantity, it is difficult to deduce the coefficient of friction throughout the contact. Consider, however, the following experiment.

A standard partial slip fretting experiment is carried out under conditions of plane deformation. Initially, the coefficient of friction is uniform over the contact, and equal to f_0. The combination of stick and slip zones is as described by the Mindlin-Cattaneo calculation (see §3.2) and is depicted in Fig. 5.16(a). After n cycles it is assumed that surface modification within the slip zones has raised the local coefficient of friction there, independently of the pressure, to f_n. At this stage, therefore, the stick-slip zone regime will be as depicted in fig. 5.16(b), but it should be emphasised that the value of the coefficient of friction, $f(x)$, is unknown in the interval $c_0 \leq |x| \leq c_n$. The stick zone size is given by:

$$\left(\frac{c_n}{a}\right)^2 = 1 - \frac{Q}{f_n P} \tag{5.47}$$

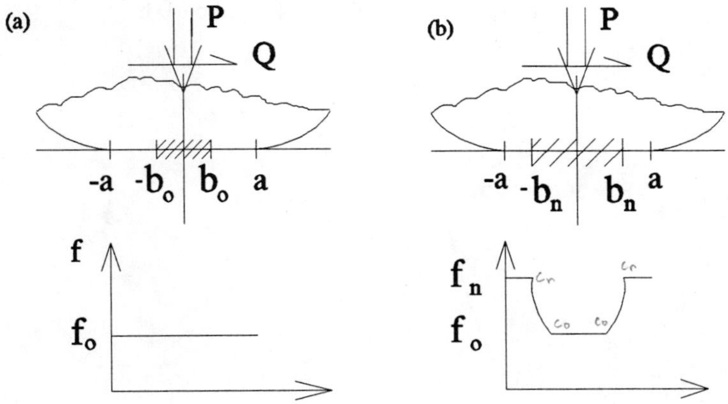

Figure 5.16: Evolution of the coefficient of friction during a partial slip fretting experiment. (a) Starting conditions, with constant initial coefficient of friction, f_0 and (b) after n cycles with current friction coefficient f_n in the slip zones.

At this point in the experiment the average coefficient of friction may be measured by stopping the fretting and then slowly increasing the tangential force until sliding occurs. The value of this measured average coefficient of friction is given by equation (5.46) i.e.,

$$\frac{\pi a \bar{f}}{4} = \int_0^{c_0} f_0 \sqrt{1-\left(\frac{x}{a}\right)^2}\, dx + \int_{c_0}^{c_n} f(x)\sqrt{1-\left(\frac{x}{a}\right)^2}\, dx + \int_{c_n}^{a} f_n \sqrt{1-\left(\frac{x}{a}\right)^2}\, dx \quad (5.48)$$

Equation (5.48) may now be differentiated with respect to the number of cycles, n, with the outcome (Hills, Nowell, and Sackfield 1988)

$$\frac{\pi a}{4}\frac{\partial \bar{f}}{\partial n} = \frac{\partial f_n}{\partial n}\left\{\frac{\pi}{4} - \frac{c_n}{2}\sqrt{1-\left(\frac{x}{a}\right)^2} - \frac{1}{2}\sin^{-1}\left(\frac{c_n}{a}\right)\right\} \quad (5.49)$$

Using equation (5.47) we find

$$\bar{f} = f_n - \frac{2}{\pi}\int_{f_0}^{f_n}\left\{\frac{Q}{f_n P}\sqrt{1-\frac{Q}{f_n P}} + \sin^{-1}\sqrt{1-\frac{Q}{f_n P}}\right\} df_n \quad (5.50)$$

The integral in this equation could clearly be evaluated, and this would then give an explicit expression giving the coefficient of friction in the slip zone, f_n, in terms of the measurable mean coefficient of friction, $\bar{f}$. In most fretting tests the initial coefficient of friction may be quite low, of the order of 0.2, so that initially sliding may be anticipated. If this is so then $f_0 = Q/P$ and equation (5.50) gives

$$\bar{f} = f_n - \frac{2Q}{\pi P}\left\{-2\sin\alpha + 2\ln\left[\tan\left(\frac{\alpha}{2}+\frac{\pi}{4}\right)\right] + \frac{P f_n \alpha}{Q} - \tan\alpha\right\} \quad (5.51)$$

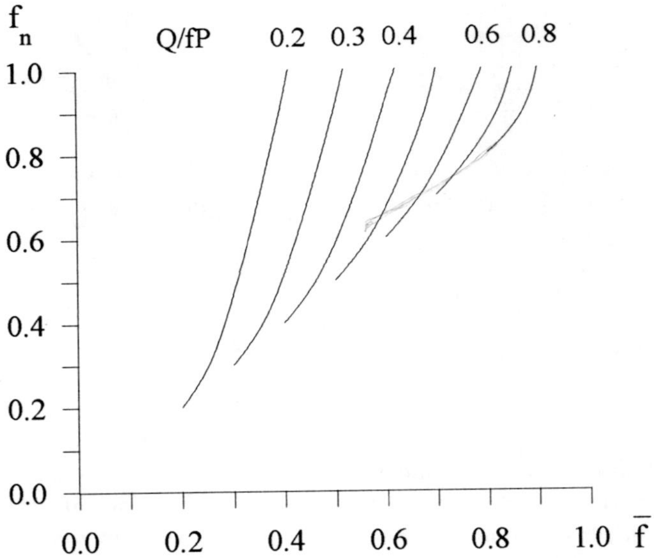

Figure 5.17: Prediction of slip zone friction coefficient, f_n, as a function of the measured mean value, $\bar{f}$, and Q/fP.

where

$$\cos\alpha = \frac{Q}{f_n P}$$

A graph of this function is shown in fig. 5.17 which therefore enables an estimate of f_n to be found from $\bar{f}$ alone, providing P, Q are both known. Typically for a Q/P ratio of 0.25 if the measured mean coefficient of friction $\bar{f}$ is 0.42 (obtained as described above by operating the apparatus momentarily under sliding conditions) the predicted value of f_n would be 0.75. This is the kind of value which might be expected to develop if the apparatus were operated under continuous sliding conditions.

It may be remarked that in the partial slip fretting fatigue tests carried out at Oxford using age-hardening aluminium alloys, the rise in true coefficient of friction with number of cycles was very rapid. Figure 5.18 indicates the evolution of coefficient of friction measured in three families of tests. Tests 1 and 2 were carried out at a peak contact pressure of 157N/mm^2, and show a rapid rise in f_n from about 0.2 to 0.55 in twenty cycles. These tests were conducted under initial sliding conditions, but after about a hundred cycles the coefficient of friction had increased to the point where partial slip obtained. Test 3 was carried out at a contact pressure of about 73 N/mm^2, and at this lower load sliding persisted to at least 150 cycles, by which time the true coefficient of friction had reached about 0.75.

In Chapter 3 the conditions for stick and slip were investigated, and it was postulated that for a pair of contacting particles within the stick zone, but about to undergo

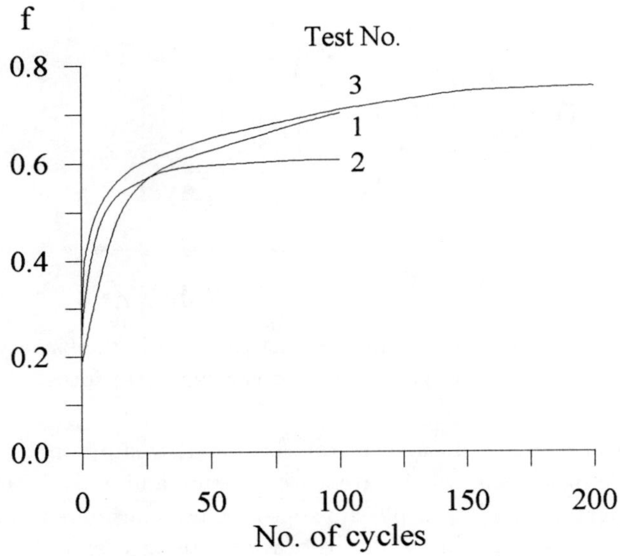

Figure 5.18: Variation of friction coefficeint with number of fretting cycles during full sliding tests on Al/4%Cu alloy.

slipping, the direction of the initial slip direction will be collinear with the resultant shearing traction direction. Thus the *increment* of slip or slip *velocity* opposes the total shearing traction. This concept is very similar to that arising in plasticity where the direction of the strain increment is usually taken to be in the same direction as the principal stress resultant. The result in plasticity is usually referred to as the 'normality' rule, since the plastic strain increment vector is normal to the yield surface. This analogy between friction and plasticity leads to the suggestion that some of the results arising in plasticity theory might be applied to friction, although in the latter case it is the resolved shear component of traction which opposes slip. In particular, where a contact undergoes cyclic loading, it is possible for regions of slip to occur during the first few cycles of loading, but for the system to shake down to a slip-free state. Thus, in a fretting contact, one way to prevent steady state slip may be to increase the normal pressure, but another may be to permit slip to occur initially but to design the contact in such a way that stress relief induced by the slipping process may inhibit further relative movement in the steady state. This is a less conservative approach, analogous to designing to the shakedown limit in elastic-plastic design. A specific example, considered by Comninou and Barber (1983) will make the process clear.

Figure 5.19 shows a layer pressed onto an elastically similar substrate, in the form of a half-plane, by a uniform, constant pressure, p_0. A shear force, Q, spatially fixed, is then increased steadily from zero and taken through a fully reversing loading

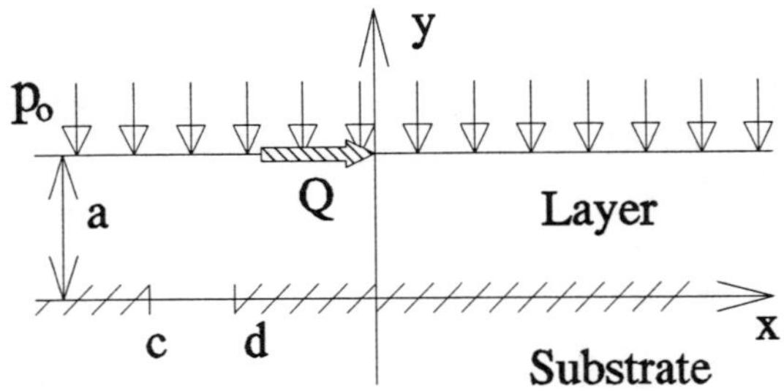

Figure 5.19: Elastic layer pressed onto an elastically similar substrate and subjected to a constant uniform pressure, p_0 and a time-varying shear force, Q.

cycle between limits $\pm Q_1$. As a result of the shear, a slip zone forms during the first quarter cycle of loading which extends between c and d. Higher values of shear which induce separation and multiple slip zones are not considered. The problem was originally solved by distributing glide dislocations along the slip zone in much the same way that they may be used for modelling cracks (see Chapter 6). For details of the calculation the original paper should be consulted, and here we will concentrate on the results found for a sample case, which are depicted in fig. 5.20. The first column shows the direct traction distribution across the interface during loading, unloading and re-loading, from $Q_1/p_0a = 2.353$, whilst the second column shows the corresponding shear tractions. The dotted lines show the extent of the slip zones. In detail, forward slip is observed to occur over the range $-1.444 < x/a < -0.400$ during the first quarter cycle of loading. During unloading, backslip occurs over the interval $0.404 < x/a < 1.439$. The backslip region is slightly smaller due to the locked-in dislocations left by the first slipping event. During re-loading, stick persists everywhere. These observations can be summarized by the table below, which maintains the analogy with elastic plastic design.

Q_1/p_0a	**Friction Problem**	**Plasticity Problem**
< 2.03	No slip at any time	Elastic limit
< 3.1	Slip in first cycle only	Shakedown limit
> 3.1	Slip in steady state	Ratchetting
	(over smaller region than first cycle)	

In a letter to the editor, Johnson (1983), suggested an analysis of the same problem, but speculatively employing the Melan shakedown theorem. Essentially, a static residual stress field $S_r(x)$ is postulated, which, together with a *reversing* shear traction distribution due to the applied shear, $S(x)$, must at all points and at all times be less than the limiting value at the onset on slip, $fN(x)$, where $N(x)$ is the resultant direction traction, i.e.,

$$|S(x) + S_r(x)| \leq fN(x) \tag{5.52}$$

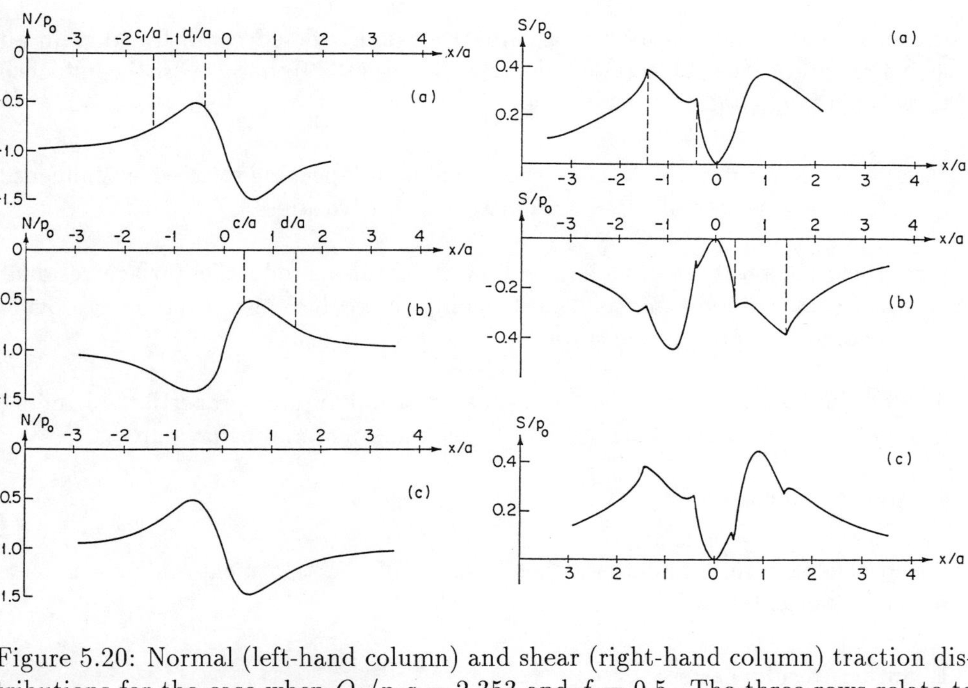

Figure 5.20: Normal (left-hand column) and shear (right-hand column) traction distributions for the case when $Q_1/p_0a = 2.353$ and $f = 0.5$. The three rows relate to (a) loading, (b) unloading, and (c) re-loading. Reproduced by permission of Int. Jnl. Solids Struct. (Pergamon Press) and courtesy of J.R. Barber.

The optimal solution in this case is when the *magnitude* of the left hand side of inequality (5.52) is on the verge of reaching the limiting value at each extreme of the loading cycle. In Johnson's analysis the influence of the distributed dislocation on the normal traction across the interface is ignored, so that $N(x)$ is given by the bulk stress field. The optimum value of the residual stress at the critical point is then found to be

$$S_r(0.826a) = 0.273p_0$$

which corresponds to

$$\frac{Q}{p_0 a} = 2f\pi = 3.14 \quad (\text{when } f = 0.5)$$

It is clear that this corresponds very closely with the value predicted by the full analysis, as shown in the table on p. 120.

5.4 The influence of surface treatments

Surface treatments may be used as a palliative against the effects of fretting. In an excellent review of the subject, Gordelier and Chivers (1979) characterise the influence of these treatments under four headings:

1. Surface mechanical treatments (shot peening, roll peening etc.) whose influence is to induce a state of surface residual compressive stress.

2. Surface chemical treatments, some of which also produce a surface residual stress (e.g. sulphidizing), but most modify the coefficient of friction or provide a means of separating the surfaces.

3. Metallic coatings which permit a change in all surface properties, including finish, residual stress, coefficient of friction and material behaviour.

4. Non-metallic coatings

From the physical standpoint, therefore, all of the treatments influence, to a greater or lesser extent:

1. Surface residual stress.

2. Coefficient of friction.

3. Surface finish.

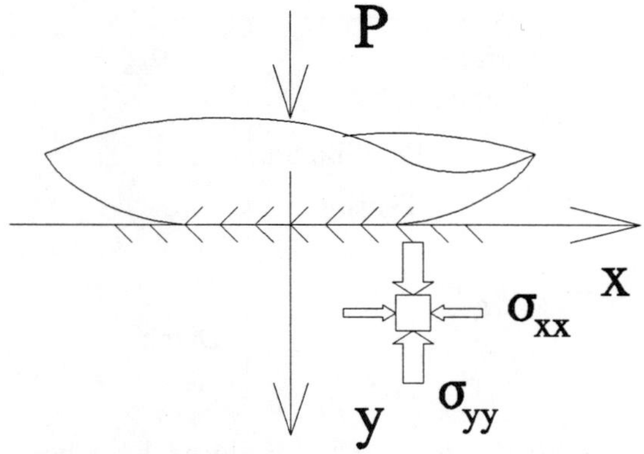

Figure 5.21: Qualitative description of the stress field induced by normal contact alone.

First, we will consider the effects of residual stress. The influence of a compressive stress parallel to the surface on crack initiation is limited, but helpful; the effect of the contact loading on near surface material is to produce a stress state where, qualitatively, the following obtains (fig. 5.21)

$$\sigma_{yy} \ll \sigma_{xx} < 0 \tag{5.53}$$

although the presence of a shearing traction may increase the value of σ_{xx} significantly, to the point where it is tensile. Crack initiation is a process controlled by the reversing shear stresses along persistent slip bands (Mura and Nakasone, 1990; Fan, Keer and Mura, 1991), which eventually lead to dislocation dipoles migrating to one end of the band, forming an extrusion. This is the embryo crack. By making the σ_{xx} component of stress more compressive, the shear stress present is reduced and hence the propelling force on the dislocations is also reduced. Thus, in problems where crack initiation is the dominant life-controlling process, this can be slowed down by the presence of residual compressive stresses, but the effect is difficult to quantify, and is, in any case, almost invariably accompanied by an increase in strength associated with strain hardening.

If a crack does start to grow, the residual stresses will also have a beneficial effect on propagation, at any rate for short cracks. A typical residual stress distribution produced by a bombardment process is sketched in fig. 5.22. This curve is typical of those produced either by shot peening or, on a much smaller scale, by ion implantation. In either case the peak residual stress may be limited by secondary plastic flow of the material (Hills, Waterhouse and Noble, 1983), and hence will be a significant proportion of the yield stress. The depth scale is characterised by the size of the contact produced by the impacting particles, and hence may be of the order of 0.5-1.0

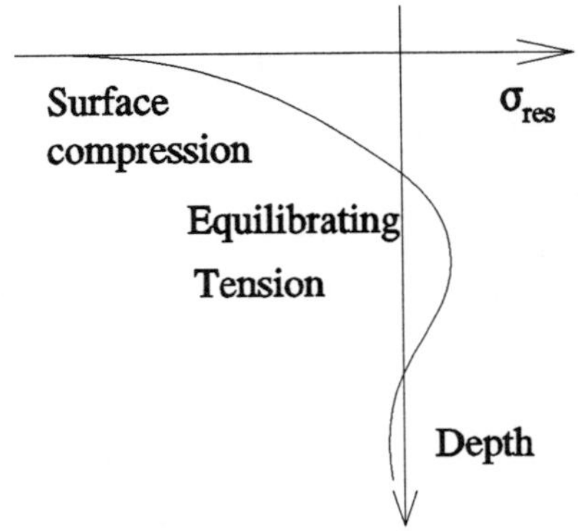

Figure 5.22: A typical residual stress profile produced by a bombardment process such as shot peening or ion implantation. The depth scale will be different for these two processes.

mm for shot-peening, but several orders of magnitude less for ion-implantation. The zone of maximum residual compressive stress is concentrated in a relatively narrow band near the surface, whilst the equilibrating tension is present over a much broader stratum deeper within the material. The compressive zone near the surface will tend to maintain closed the faces of a normal crack growing down from the free surface. Thus, if the sum of all the direct stresses acting in the x-direction and present near the surface (residual stresses, contact stresses, any bulk applied tension) is always negative, there is little possibility of a crack developing. If, despite the presence of the residual stresses, crack opening *does* occur, growth may be expected inwards towards the region of tension. If the crack breaks through here it will suddenly experience significant additional crack tip stress intensification, and accelerated growth. Also, as the presence of the surface compression does so much to hinder crack development from a point of surface initiation, initiation may occur at a subsurface point, typically starting from some stress-raising feature such as an inclusion. If this happens, the crack will grow both deeper and towards the surface until, fig. 5.23, the stress intensity factor range at the near-surface end falls to zero (or below the threshold value) so that the crack tip closes. The analysis of this configuration by using the method of distributed dislocations is described in §6.3. It should be mentioned that it is also possible to analyse cracks in this configuration by using the finite element method (Wilks, Nowell and Hills, 1993), but considerable care is needed.

If the finite element method is employed to analyse residual stress problems, it is important to understand that the residual stress distribution cannot be represented merely as applied loads over the ends of the region (see §6.3). This is because field equilibrium requirements mean that the variation in residual stress with depth will gradually diminish remote from the edges, leaving just the mean stress which, for a

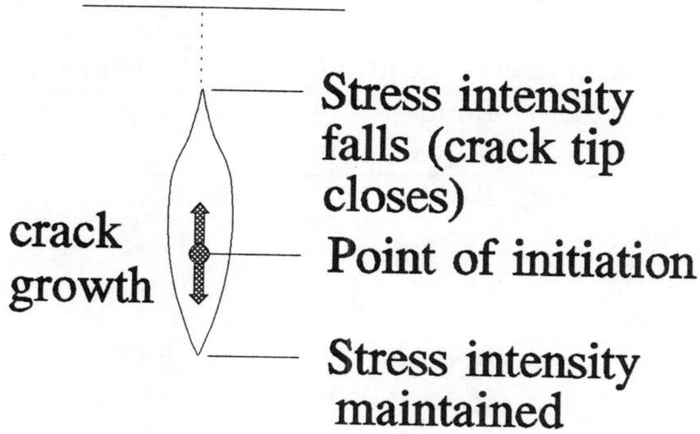

Figure 5.23: The form of crack developed when a subsurface flaw grows in a residual stress field of the type depicted in fig. 5.22.

self-equilibrating load, will be zero. Therefore, each element must have assigned to it a pre-stress at the start of the analysis equal to the residual stress there, but with no associated strain. If this procedure is followed, and special contact elements are used to model crack flank closure, reliable results may be obtained.

Let us now examine the influence of friction on fretting fatigue. Chivers and Gordelier (1985) followed up their earlier review by a detailed consideration of how changes in the coefficient of friction influence the predicted fretting fatigue performance. Their initial assumption is that crack initiation is controlled by the maximum *tensile stress* developed at the surface[4]. They go on to investigate whether palliatives which are designed to modify the coefficient of friction ought to act so as to increase or decrease its value. They study both plane and axi-symmetric Hertzian contacts under conditions of partial slip, and use the Mindlin-Cattaneo solution to model the surface shear traction. Further, in their solution they consider two kinds of tangential loading, viz. imposed displacement conditions or imposed force conditions. Lastly, they consider the influence of modifying the normal load. Their conclusions are summarized in the table overleaf. The apparently conflicting influences of changing these two basic quantities indicate the complexity of contacts under partial slip conditions.

[4]The component of tensile stress considered is that parallel to the free surface. Note that for points lying outside the contact (or at its edge) this is equivalent to the maximum shear stress, since no normal or shear tractions are present exterior to the contact.

	Axi-symmetric contact		**Plane contact**	
Contact condition	Imposed force	Imposed displacement	Imposed force	Imposed displacement
Coefficient of friction	Increase (at low *f*) Reduce (at high *f*)	Reduce	Reduce	Reduce
Normal contact load	Increase	Reduce (near full slip) Increase (near zero slip)	No effect	Reduce

Recommendations for altering the coefficient of friction in order to improve fretting fatigue life in the partial slip regime (from Chivers and Gordelier, 1985).

Chapter 6

The Analysis of Cracks

6.1 Introduction

The object of studies of fretting is ultimately to be able to quantify the rate of fatigue crack growth, and to be able to describe the conditions under which cracks do not arise, or at any rate do not grow. Traditional approaches to the quantification of fatigue have relied on the assumption that the nominal stress in the absence of the crack is sufficient to determine its behaviour. The classical experiments of Wöhler were designed to measure the fatigue life of railway axles under fully reversing stress, which was calculated from elementary bending theory. A graph of the number of cycles to failure against the bending stress became known as an S-N curve (Dieter, 1961), and subsequent experimental work has enabled the influence of mean stress to be taken into account. However, a bulk approach to fatigue under a multi-axial state of stress has never been fully developed (Toor, 1975 a,b). Further, the initiation and propagation elements of fatigue crack growth cannot be separated by this approach.

A more modern technique for analyzing cracks, including crack growth rate, is to use fracture mechanics. It is not possible here to provide a thorough introduction, for which the recent texts by Gdoutos (1990, 1993), Ewalds and Wanhill (1984), Broek (1982, 1989) or Derby, Hills and Ruiz (1992) might be consulted. We will, however, give some justification for the Paris fatigue crack propagation law.

It is well known that a sharp external corner in an arbitrarily loaded component is free of stress while an internal corner suffers a stress singularity. A systematic technique for explaining this phenomenon was devised by Williams (1952), who investigated the state of stress near the apex of a wedge. Williams argued that if a singular state of stress existed, its characteristics would be geometry dependent, but would not be unduly influenced by the remote loading. Accordingly, he chose to represent the state of stress near the apex as a harmonic series in the r-direction, and with an r-direction variation of the form r^λ. The value of λ was found to be positive (corresponding to a bounded state of stress as $r \to 0$) for wedge angles less than 180° but was increasingly negative for re-entrant wedges with larger included

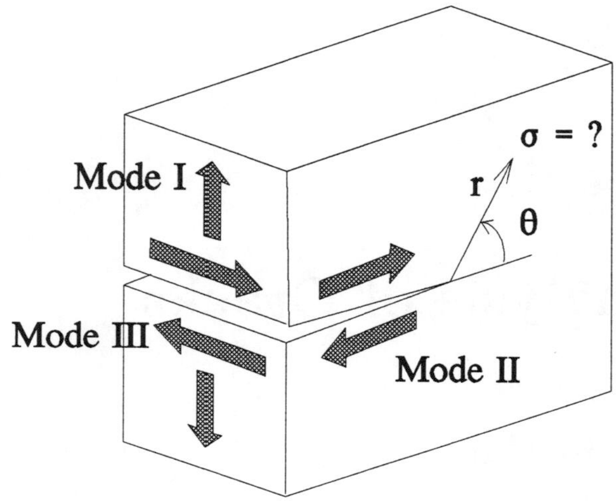

Figure 6.1: The three modes of loading of the crack tip.

angles. In the limit, as the wedge angle is increased to $360° - \epsilon$, the value of λ becomes $-\frac{1}{2}$. Thus, it can be seen that the state of stress near a crack tip in a homogeneous material will always be square root singular. Furthermore, solution of the state of stress for a crack in an infinite plate shows that the θ-variation of stress can have one of two characteristic forms (Westergaard, 1939) depending on whether the far field is tension or shear. In addition, a third form of loading can arise if the applied loads are transverse to the plate, inducing a through-thickness shear; this, too, has a characteristic near crack-tip stress field. It follows that the state of stress near a crack tip may be written in the form

$$\sigma_{ij}(r,\theta) = \frac{1}{\sqrt{2\pi r}}\left\{K_I f_{ij}^I(\theta) + K_{II} f_{ij}^{II}(\theta) + K_{III} f_{ij}^{III}(\theta)\right\} \tag{6.1}$$

where the three modes of loading I, II, III are defined in fig. 6.1, and the quantities K_I, K_{II}, K_{III} are known as modes I, II and III stress intensity factors respectively. If the stress components are given in polar coordinates, the functions $f_{ij}(\theta)$ are defined by the following three sets of equations.

$$\left\{\begin{array}{c} f_{rr}^I \\ f_{\theta\theta}^I \\ f_{r\theta}^I \end{array}\right\} = \left\{\begin{array}{c} \cos(\theta/2)\,[1 + \sin^2(\theta/2)] \\ \cos^3(\theta/2) \\ \sin(\theta/2)\cos^2(\theta/2) \end{array}\right\} \tag{6.2}$$

$$\left\{\begin{array}{c} f_{rr}^{II} \\ f_{\theta\theta}^{II} \\ f_{r\theta}^{II} \end{array}\right\} = \left\{\begin{array}{c} \sin(\theta/2)\,[1 - 3\sin^2(\theta/2)] \\ -3\sin(\theta/2)\cos^2(\theta/2) \\ \cos(\theta/2)[1 - 3\sin^2(\theta/2)] \end{array}\right\} \tag{6.3}$$

$$\left\{ \begin{array}{c} f_{rz}^{III} \\ f_{\theta z}^{III} \end{array} \right\} = \left\{ \begin{array}{c} \sin(\theta/2) \\ \cos(\theta/2) \end{array} \right\} \tag{6.4}$$

Components of stress other than those quoted are zero, except σ_{zz}which is determined by the degree of transverse constraint (plane stress or plane strain).

The components of stress may be substituted into Hooke's Law to obtain the strains, and these in turn may be integrated to give the displacement field. The displacements may be written as

$$u_i(r,\theta) = \frac{1}{2E}\sqrt{\frac{r}{2\pi}}\left\{K_I g_i^I(\theta) + K_{II} g_i^{II}(\theta) + K_{III} g_i^{III}(\theta)\right\} \tag{6.5}$$

which indicates that all the displacement components vary as $\sqrt{r}$ and the functions are

$$\left\{ \begin{array}{c} g_r^I \\ g_\theta^I \end{array} \right\} = \left\{ \begin{array}{c} (1+\nu)[(2\kappa-1)\cos(\theta/2) - \cos(3\theta/2)] \\ (1+v\nu)[-(2\kappa+1)\sin(\theta/2) + \sin(3\theta/2)] \end{array} \right\} \tag{6.6}$$

$$\left\{ \begin{array}{c} g_r^{II} \\ g_\theta^{II} \end{array} \right\} = \left\{ \begin{array}{c} (1+\nu)[-(2\kappa-1)\sin(\theta/2) + 3\sin(3\theta/2)] \\ (1+\nu)[-(2\kappa+1)\cos(\theta/2) + 3\cos(3\theta/2)] \end{array} \right\} \tag{6.7}$$

$$g_z^{III} = 2(1+\nu)\sin(\theta/2) \tag{6.8}$$

Here, ν is Poisson's ratio, $\kappa = (3-v)/(1+\nu)$ under plane stress conditions and $3-4\nu$ under plane strain. It should be emphasized that stresses (6.1) arise from taking the dominant singular term in a series expansion. Bounded terms for the stress components are also present, but will be swamped by the singular term if r is small enough.

A detailed discussion of the stress intensity factor and its determination is given in §6.2. However, for the time being it is useful to note that it can often be described by an equation of the form

$$K_I = Y\sigma_0\sqrt{\pi b} \tag{6.9}$$

where σ_0 is a representative applied stress, b is the crack half-length and Y is a function representing the influence of the geometry. For example, for a crack in a remote tension field with distant boundaries, $Y \equiv 1$.

The equations derived above are of the utmost importance in the study of brittle fracture, whether through monotonic loading or fatigue. It may be shown (e.g. Derby, Ruiz and Hills, 1992) that the three stress intensity factors are related to the generalized crack extension force or strain energy release rate, G, by

$$G = \frac{1}{8\mu}\left\{(1+\kappa)(K_I^2 + K_{II}^2) + 4K_{III}^2\right\} \tag{6.10}$$

where μ is the modulus of rigidity. The quantity G is the crack propelling term as defined by Griffith's criterion (Ewalds and Wanhill, 1986), and hence a necessary condition for crack growth on energetic grounds is that it should reach a critical value. However, energetic considerations of crack extension are mainly useful where the material being fractured is almost infinitely elastic, i.e. there is completely brittle fracture. Their main use is in the basic consideration of the fracture of glass and possibly ceramics. The stress intensity factor encapsulates much more information than the generalized crack extension force; it gives the magnitude of the state of stress, strain and displacement ahead of the crack in terms of the far-field loading, whilst the spatial distribution of these quantities is given by equations (6.2) to (6.8). Thus, all of the relevant quantities which might have a bearing on crack extension are fully characterised, and the fracture criterion under any mode of loading, providing that crack-tip plasticity is confined to a region where the surrounding elastic material is well within the region dominated by the singular field, may be written as

$$K_i = K_{iC} \qquad i = I, II, III \tag{6.11}$$

where the quantities K_{iC} are material properties known as fracture toughnesses. It is difficult to develop such a convincing argument under combined mode loading, as it is not clear what functions or combinations of K_I, K_{II}, K_{III} will give rise to a critical crack tip condition for crack extension, and it is customary to revert to energy considerations, so that crack extension is assumed to occur when G reaches a critical value.

The elegance of the fracture mechanics approach is that the same arguments relating to crack extension under monotonic loading will also apply to fatigue. It is generally accepted that irreversibilities in the crack tip plastic zone give rise to incremental crack extension, and that since the whole of the characteristics of the plastic field are controlled by the surrounding elastic material, they, too, may be described using the singular solutions and stress intensity factors. We do not need to enquire into the details of the crack extension process to state that growth rate must be described by an equation of the form

$$\frac{db}{dN} = f(\Delta K) \tag{6.12}$$

where b is crack length, N is number of cycles and now ΔK is the range of stress intensity factor the crack tip suffers. The form of the function can only be found experimentally. Fatigue tests are carried out using a range of geometries of cracks, but having the same stress intensity factor, and it is found that the crack growth rate depends on ΔK alone. If the results of a large number of tests are plotted on a logarithmic diagram, the pattern of fig. 6.2 emerges. Below a certain stress intensity factor, ΔK_{th}, the so called threshold value, no crack growth occurs[1]. This corresponds to the notion of a fatigue limit in classical fatigue (Dieter, 1961). At slightly higher stress intensities the crack growth rate accelerates significantly, and this is said to be stage 1 growth (Forsyth, 1961). During this phase the crack is still

[1]The existence of a threshold value is the subject of some controversy. If the low ΔK is the result of a very short crack but high loads, significant local plasticity may occur and the crack may be in the process of undergoing initiation.

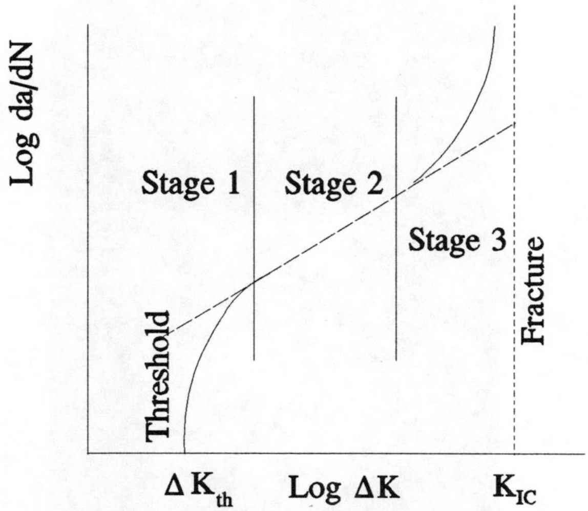

Figure 6.2: Crack growth as a function of ΔK .

normally quite short and is making the transition from the initiation phase, which, for fatigue of precision components, normally involves very localized plasticity and the ratchetting of dislocations along persistent slip bands (Mura and Nakasone 1990), to the development of a proper crack straddling more than one grain. Stage 1 growth is associated, therefore, with phenomena controlled by shear (plasticity and the glide of dislocations (Fan, Keer and Mura, 1991)), so that it is usually found that stage 1 cracks grow along planes of maximum shear (or near maximum if the preferred glide planes are not oriented ideally), which for surface cracks lie at $\pm 45°$ to the free surface. As the stress intensity factor increases the linear portion of fig. 6.2 is entered. This transition from stage 1 to stage 2 is marked by a change in the physical orientation of the crack tip, which kinks so that the crack tip is subjected to mode I loading only. A typical example is shown in fig. 6.3, which is a failed fretting fatigue specimen. Stage 2 growth normally takes up a considerable proportion of the crack's life, and equation (6.12) for this regime takes on a simple form, viz.

$$\frac{db}{dN} = A(\Delta K)^m \tag{6.13}$$

in which form it is known as Paris' equation (Paris, Gomez, and Anderson, 1961). This function is so widely used that the constants A, m are tabulated for many materials (Derby, Hills and Ruiz, 1992). Eventually the crack becomes so long that the crack tip plasticity zone starts to dominate the response and in his third stage of growth acceleration towards failure (as $\Delta K \to K_{IC}$) occurs.

A typical loading history is depicted in fig. 6.4. This emphasises the fact that two

Figure 6.3: Photograph of a failed fretting fatigue specimen showing stages 1 and 2 growth.

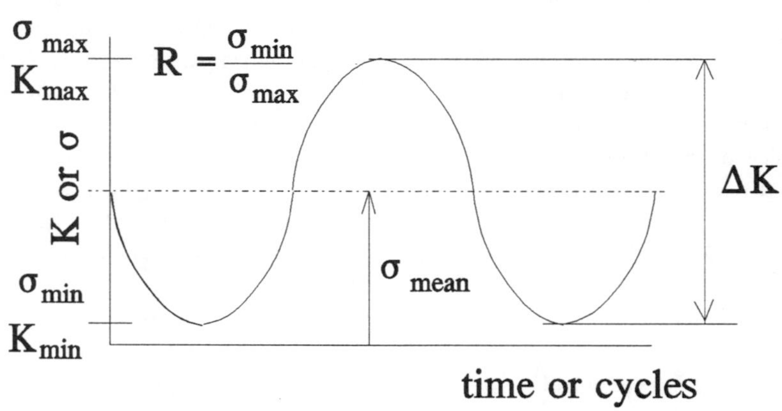

Figure 6.4: Stress ratio; two quantities are needed to describe the stress or stress intensity history in general.

quantities are needed to describe the amplitude of the waveform seen by the crack tip. For example, if we denote the maximum and minimum stress intensity factors by $K_I^{\max}, K_I^{\min}$ respectively the range of stress intensity factor employed above is given by

$$\begin{aligned} \Delta K &= K_I^{\max} - K_I^{\min} \\ &= K_I(\sigma^{\max} - \sigma^{\min}; b) \end{aligned} \tag{6.14}$$

where $\sigma^{\max}, \sigma^{\min}$ represent the maximum and minimum stresses used in the definition of the stress intensity factor (equation 6.9), and b is the crack half-length. The rate of crack growth is strongly dependent on this quantity, but it is also more weakly dependent on the mean stress present. It is customary to use not the *mean* stress intensity factor as the second loading variable, but the stress ratio, defined by

$$R = \frac{\sigma^{\min}}{\sigma^{\max}} = \frac{K^{\min}}{K^{\max}} \tag{6.15}$$

and materials data both for the threshold, ΔK_{th}, and Paris' Law constants (A, m) may be tabulated in terms of R.

In most structures, cracks grow in an environment in which the overall character of the stress field is tensile; that is, the crack is normally held open for much of its life, and it invariably grows in such a manner that in stage 2 the crack tip stress intensity factor is purely mode I - in other words the plane of the crack is perpendicular to the greatest positive principal stress. Under these conditions, and ignoring for the time being any localized crack tip plasticity, the range of stress intensity factor experienced by the crack tip will be given by equation (6.14) where $\sigma^{\min}$ is the minimum remote stress actually experienced or zero, whichever is greater. The reason for stating that the minimum possible value of the stress intensity factor is zero arises from purely physical reasoning: in tension the crack faces can support no direct tractions, and hence separate, giving rise to the stress intensity described earlier. But in *compression* the crack faces merely contact each other and transmit a contact pressure, so that the presence of the crack is ignored. There is therefore no stress intensity at the crack tip.

In the case of *uniform* compression described above crack closure occurs along the whole length of the crack, as it was assumed that there was global compression perpendicular to the plane of the crack. However, in cracks propelled by contact loading, for example in a crack initiated by fretting fatigue, partial closure may arise, fig. 6.5. This is because the stress gradients in the neighbourhood of the crack are very severe, and whereas part of the crack might be in a tensile region another part might be in compression. A simple practical example of this is given by a crack normal to a shot-peened surface. The residual stress profile is normally compressive at the surface, but with a zone of tension at greater depths, necessary to ensure equilibrium (fig. 6.5). Here, the crack faces at the surface remain pressed together whilst those

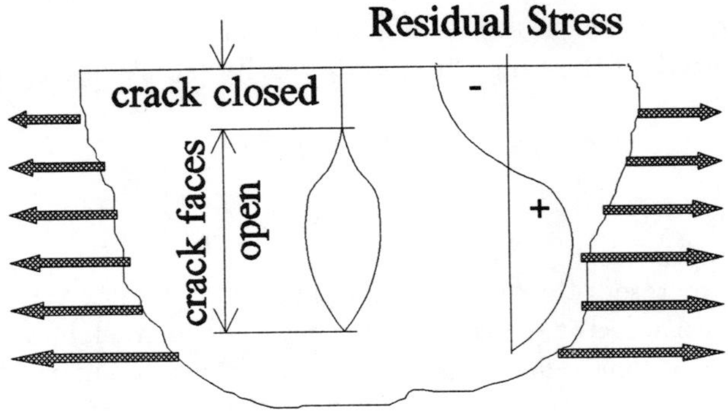

Figure 6.5: Partial closure of a crack, in this case caused by the rapidly varying residual stress profile.

near the crack tip will separate, giving rise to a finite stress intensity factor. Great care is needed in calculating the crack tip stress intensity under these conditions.

These comments relating to crack closure or partial closure have been made within the context of elasticity theory. Another source of crack tip closure is the result of relaxation of the crack-tip plastic zone during unloading. A detailed discussion of this phenomenon is outside the scope of this book but may be found in Kanninen and Popelar (1985), *inter alia.*

It will be apparent from this resumé of fracture mechanics that the quantity playing the key role in driving cracks is the crack tip stress intensity, K. In the next section we shall review the various methods employed to find this quantity, and proceed to describe in some detail a technique ideally suited to this purpose in the context of cracks propelled by contacts, and possible partial crack closure.

6.2 Analysis of plane cracks

We have seen that the key quantity controlling both crack growth rate and fracture is the stress intensity factor. Its determination is therefore of the utmost importance, and, whilst all cracks are in practice to some degree three dimensional, a great deal of insight can be obtained by idealizing them by a two-dimensional geometry. Under plane conditions it is possible to find the stress intensities for a very wide range of geometries with some rigour, even if the crack exists in a steep stress gradient, as is frequently the case when the crack is near a contact.

First, the possible methods of solution available will be reviewed. The early solutions found, from which a knowledge of the crack tip asymptotics was inferred, were due to Westergaard (1939), for a crack in an infinite medium. They were obtained by determining the Airy stress function, and are therefore exact within the assumptions

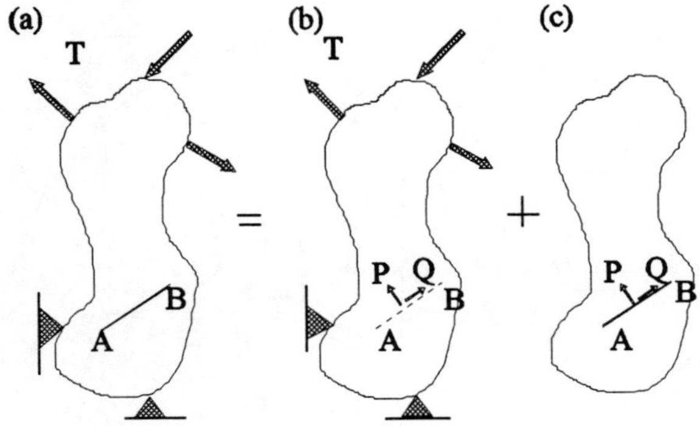

Figure 6.6: Bueckner's Principle; (a) is the cracked, externally loaded body to be analysed, (b) the perfect body under the same external loading and (c) the influence of equilibrating tractions acting along the line of the crack alone.

of linear elasticity. Unfortunately, although it is possible to devise stress function solutions for more complex loading (Muskhelishvili, 1953), it is very difficult to take remote boundaries into account. A series stress function solution (Williams, 1957) in principle makes this possible, but boundary conditions are only approximated by the method. A more promising technique is to use Green's functions for crack forces in combination with the Bueckner superposition principle, and as this is a useful strategy we shall describe it before passing on.

Bueckner's principle (Bueckner, 1958) may be appreciated by reference to fig. 6.6. Figure 6.6(a) shows the cracked, loaded body for which we would like to find the crack tip stress intensity factors $K_I^A, K_I^B, K_{II}^A, K_{II}^B$. The first step in the solution is to find the state of stress which would occur if the crack were absent. In particular, we require, the traction components $\sigma_{yy}(x), \tau_{xy}(x)$, along the line of the crack, fig. 6.6(b). Separately, we consider the same body, devoid of external loading, but with tractions $-\sigma_{yy}(x), -\tau_{xy}(x)$, acting along the line of the crack, fig. 6.6(c). If their contributions to the crack tip stress intensities were known, the problem would be solved, as the sum of the solutions shown in fig. 6.6(b) and (c) is equivalent to the original problem. Unfortunately, the only Green's functions of this type known in closed form are for a crack in an infinite body (Irwin, 1957). Figure 6.7 shows a crack subject to opening and shearing forces, P and Q respectively which induce stress intensity factors at end B of

$$K_I^B = \frac{1}{2\sqrt{\pi b}}\left\{Q\left(\frac{\kappa-1}{\kappa+1}\right) + P\sqrt{\frac{b+\xi}{b-\xi}}\right\} \tag{6.16}$$

$$K_{II}^B = \frac{1}{2\sqrt{\pi b}}\left\{P\left(\frac{\kappa-1}{\kappa+1}\right) + Q\sqrt{\frac{b+\xi}{b-\xi}}\right\} \tag{6.17}$$

where $\kappa = (3-\nu)/(1+\nu)$ in plane stress or $(3-4\nu)$ in plane strain. For equal

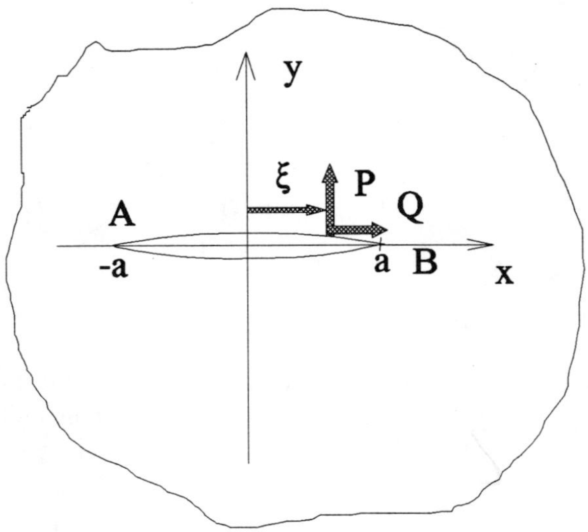

Figure 6.7: Green's functions for an infinite body.

and opposite forces (shown dashed in fig. 6.7) the coupling terms vanish, and for distributed tractions we have

$$K_I^B = \frac{1}{\sqrt{\pi b}} \int_{-b}^{b} \sigma_{yy}(\xi) \sqrt{\frac{b+\xi}{b-\xi}} d\xi \tag{6.18}$$

$$K_{II}^B = \frac{1}{\sqrt{\pi b}} \int_{-b}^{b} \tau_{xy}(\xi) \sqrt{\frac{b+\xi}{b-\xi}} d\xi \tag{6.19}$$

Recently numerical methods have become popular. Foremost is the finite element method. Care must be taken to use special elements around the crack tip to enable the $1/\sqrt{r}$ singularity to be accommodated, and care must also be taken in abstracting the stress intensity factors. The most popular technique is probably to determine the so-called J-integral (Ewalds and Wanhill, 1986), as this can be done automatically, but as this is equivalent to the strain energy release rate, G, under fully elastic conditions, it can yield only $K_I^2 + K_{II}^2$ (equation 6.10) in a plane analysis, and is incapable of separating the modes of loading. A more informative, although time consuming method is to note the asymptotic form of the stress for small r (equation 6.1), and hence a graph of, for example, $\sigma_{yy}\sqrt{2\pi r}$ against r will yield, in the limit $r \rightarrow 0$ the value of K_I.

Another general approach used with some success is the boundary element method (Rooke, Rayaprolu and Aliabadi, 1992, Aliabadi, Rooke and Cartwright, 1987), which is described in detail by Aliabadi and Rooke (1991) in the context of crack problems.

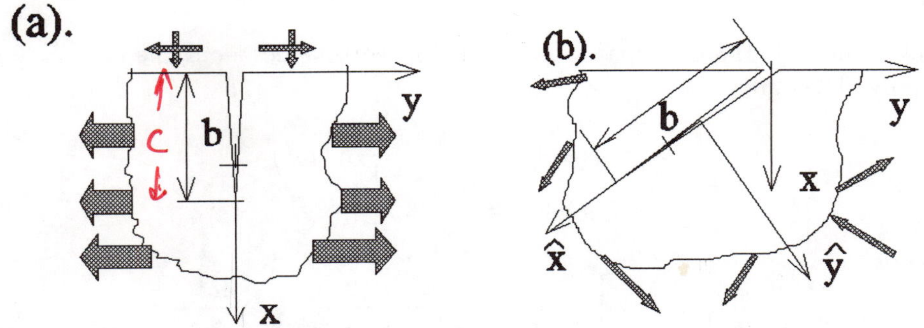

Figure 6.8: A half plane subjected to arbitrary loading and containing (a) a normal crack, and (b) an inclined crack

It is clear that there are many potential methods available, and most of the above are possible contenders. However, two features of the crack in the neighbourhood of a contact mean that considerable care is needed in the choice, viz.

(i) the cracks invariably exist in a steep stress gradient.

(ii) the cracks may be partially closed, which rules out the use of most methods relying on superposition, unless particular care is exercised in ensuring that the unilateral boundary conditions needed are satisfied. These will be discussed in detail in §6.3.

The Green's function approach described above seems attractive, but unfortunately it is very difficult to find the stress field for a crack in a half-plane loaded by a force-pair. An alternative, which at first sight is far less intuitive, but on further investigation will be seen to be extremely powerful, is to distribute displacement discontinuities or dislocations along the line of the crack. It should be emphasised here that although these entities have exactly the same properties as crystallographic defects of the same name they are here nothing more than a device for introducing a state of self-stress; no lattice defects are implied.

The first problem we will consider is shown in fig. 6.8(a). A half plane ($x > 0$) is subjected to an arbitrary loading, but such that the crack depicted suffers only mode I loading; this is achieved here by imposing symmetrical loading. The first step in the solution is to find the stress, $\sigma_T(x)$, along the line of the crack ($y = 0, 0 \le x \le b$) in its absence. We now invoke Bueckners' theorem and distribute dislocations along the line of the crack so that the stresses induced cancel $\sigma_T(x)$. To do this let us first examine the characteristics of a single dislocation installed at a point $(c, 0)$.

A dislocation is formed by cutting out a narrow (but finite) strip of width b_y between the point $(0, 0)$ and $(c, 0)$ and glueing the faces of the cuts together along their entire length. This introduces a displacement discontinuity of magnitude b_y (the Burgers vector) at point $(c, 0)$. The state of stress induced by the dislocation within a half-plane has been found by Dundurs and Mura (1964). Using results given by Dundurs and Sendeckyj (1965) we may write down the stress values along the line $y = 0$ very simply. They are

$$\begin{aligned} \widetilde{\sigma}_{yy}(x,0) &= \frac{\mu b_y(c)}{\pi(\kappa+1)} K(x,c) \\ \widetilde{\tau}_{xy}(x,0) &= 0 \end{aligned} \tag{6.20}$$

where

$$K(x,c) = 2\left\{\frac{1}{x-c} - \frac{1}{x+c} - \frac{2c}{(x+c)^2} + \frac{4c^2}{(x+c)^3}\right\} \tag{6.21}$$

If, now, instead of installing a discrete dislocation, we introduce a *distribution* of dislocations of density $B_y(c)$ where

$$B_y(c) = \frac{db_y}{dx}(c) \tag{6.22}$$

we can see that the total normal traction across the crack, $N(x)$, is given by

$$N(x) = \sigma_T(x) + \frac{\mu}{\pi(\kappa+1)} \int_0^b B_y(c) K(x,c) dc. \tag{6.23}$$

In order to render the crack faces traction free we set

$$N(x) = 0 \qquad 0 \le x \le b \tag{6.24}$$

This gives an integral equation in the unknown function $B_y(c)$; it has a so-called generalized Cauchy kernel, as the remainder of the integrand, $K(x,c)$, contains bounded terms and a singular term, viz. $1/(x-c)$. Once again the numerical integration scheme described in §4.4.1 may be employed to determine $B_y(c)$. The first step is to re-write the equation over a normalized interval by substituting

$$\begin{aligned} r &= 2c/b - 1 \\ s &= 2x/b - 1 \end{aligned} \tag{6.25}$$

giving

$$\frac{\mu}{\pi(\kappa+1)} \int_{-1}^1 B_y(r) K(s,r) dr = -\sigma_T(s) \qquad -1 \le s \le 1 \tag{6.26}$$

Now, the quantity $B_y(r)$ has a physical meaning: it represents (equation 6.22) the angle between the crack faces, or their relative slope. From equation (6.5), but using the coordinates of fig. 6.8(a) we can see that the displacement of the crack faces from their unloaded position near the crack tip is of the form

$$u_\theta \propto K_I \sqrt{b-x} \tag{6.27}$$

so that the angle between the crack faces varies like

$$\frac{du_\theta}{dx} \propto \frac{K_I}{\sqrt{b-x}} \tag{6.28}$$

On the other hand as $x \to 0$ we expect the crack faces to be nearly, but not precisely, parallel. We can incorporate this required behaviour of $B_y(r)$ into the solution approximately by writing[2]

$$B_y(r) = \phi(r)\sqrt{\frac{1+r}{1-r}} \tag{6.29}$$

so that the unknown function $\phi(r)$ is smooth, and regular. Substituting (6.29) into (6.26) permits the special Gauss-Jacobi quadrature as described by Erdogan, Gupta and Cook (1973) (see §4.4.1). Both the Cauchy-singular nature of the integrand and the behaviour at each of the interval are incorporated, and the discretised form of equation (6.26) is

$$\frac{\mu}{\pi(\kappa+1)}\sum_{i=1}^{n}\frac{2\pi(1+r_i)}{2n+1}K(s_k,r_i)\phi(r_i) = -\sigma_T(s_k) \tag{6.30}$$

where

$$r_i = \cos\left(\frac{2i-1}{2n+1}\pi\right) \qquad i = 1,2,...,n$$

$$s_k = \cos\left(\frac{2k}{2n+1}\pi\right) \qquad k = 1,2,...,n$$

Thus, we have a system of n simultaneous linear algebraic equations for the n unknowns $\phi(r_i)$. They may readily be solved using any standard procedure, and, via equation (6.29) permit the shape of the crack faces to be deduced at any point. More importantly, the crack tip stress intensity factor may be found directly from a combination of equations (6.30, 6.28 and 6.29). It is observed that the stress intensity factor is proportional to $\phi(1)$ and it is found that

$$K_I = 2\sqrt{2}\sqrt{\pi b}\frac{\mu}{(\kappa+1)}\phi(1) \tag{6.31}$$

It will be recognised that the value of ϕ is revealed accurately only at the integration points, r_i, and that for its determination at other points we need to examine the definition of the underlying Jacobi interpolation polynomials. From this we can deduce formulae for the evaluation of ϕ at any point; this has been done by Krenk (1975), who showed that

$$\phi(1) = \frac{2}{2n+1}\sum_{i=1}^{n}\cot\left(\frac{2i-1}{2n+1}\frac{\pi}{2}\right)\sin\left(\frac{n}{2n+1}(2i-1)\pi\right)\phi(r_i) \tag{6.32}$$

Thus, a precise and rapid technique for the evaluation of the stress intensity factor has been developed. It may be remarked that an entirely analogous set of equations

[2] It will be recognised that this assumed form forces the crack faces to be *parallel* as $x \to 0$. A refinement of the method permitting a divergence of the crack faces has been described by Dewynne, Hills and Nowell (1992) but it was shown that the influence on the crack tip stress intensity is small.

may be developed for a geometrically similar crack loaded in shear. For example, the influence of a dislocation with Burgers vector $b_x(c)$ is to induce stresses

$$\begin{aligned} \tilde{\sigma}_{yy}(x,0) &= 0 \\ \tilde{\tau}_{xy}(x,0) &= \frac{\mu b_x(c)}{\pi(\kappa+1)} K(x,c) \end{aligned} \tag{6.33}$$

where $K(x,c)$ is still given by (6.21). A dislocation density is defined in a similar way and we find, in lieu of equation (6.26), that

$$\frac{\mu}{\pi(\kappa+1)} \int_{-1}^{1} B_x(r) K(r,s) dr = -\tau_T(s) \qquad -1 \leq s \leq 1 \tag{6.34}$$

where $\tau_T(s)$ is the shear traction appearing on the line of the crack due to the remote loading. The discretised form of equation (6.34) is given by equation (6.30) *mutatis mutandis,* and the mode II stress intensity is similarly given by equation 6.31, with K_I replaced by K_{II}.

In assuming that the crack is traction-free (equation 6.24) we have implicitly assumed that it is open throughout its length. This may be verified *a posteriori* by making use of the definition of the dislocation density (equation 6.22). As the tip is closed the opening displacement at any point, ξ is given by

$$\Delta(\xi) = \int_{\xi}^{b} B_y(x) dx \tag{6.35}$$

It is normally sufficient to determine the value of $B_y(x)$ at the integration points r_i, and to use the trapezium integration rule to verify that

$$\Delta(\xi) \geq 0 \qquad 0 \leq \xi \leq b \tag{6.36}$$

Usually when a crack initiates from a nominally flawless cyclically loaded region it grows initially in mode II (see §6.1) and hence the geometry depicted in fig. 6.8(b) may arise. Slanting cracks produce several additional complications which we will now address. The first is that both climb ($b_{\hat{y}}$) and glide ($b_{\hat{x}}$) dislocations induce both direct and shear tractions along the line of the crack ($\hat{y} = 0$) so that there is coupling between the mode I and II loadings. The starting point of the calculation is to find the state of stress induced by the dislocations, and this is provided by the Dundurs and Mura solution (1964). For a dislocation having Burgers vector components (b_x, b_y) located at $(c, 0)$ in the global axis set of fig. 6.8(b), the state of stress induced is given by

$$\left\{ \begin{array}{c} \sigma_{xx} \\ \sigma_{yy} \\ \tau_{xy} \end{array} \right\} = \frac{\mu}{\pi(\kappa+1)} \left\{ \left\{ \begin{array}{c} G_{xxx}(x,y,c) \\ G_{xyy}(x,y,c) \\ G_{xxy}(x,y,c) \end{array} \right\} b_x + \left\{ \begin{array}{c} G_{yxx}(x,y,c) \\ G_{yyy}(x,y,c) \\ G_{yxy}(x,y,c) \end{array} \right\} b_y \right\} \tag{6.37}$$

where the functions defined, G_{ijk} are given explicitly in Appendix A. Of course these equations relate Burgers vectors expressed in global (x, y) coordinates to stress components expressed in global coordinates. In order to write down an equation expressing the requirement that the tractions along the crack faces vanish, we need to transform the stresses into a local set of coordinates, fig. 6.8(b). Also, although it is not essential for open cracks, we may also express the Burgers vector components in local coordinates. Thus

$$\begin{aligned} \tau_{\widehat{xy}} &= -\tfrac{1}{2}(\sigma_{xx} - \sigma_{yy}) \sin 2\theta + \tau_{xy} \cos 2\theta \\ \sigma_{\widehat{yy}} &= \tfrac{1}{2}(\sigma_{xx} + \sigma_{yy}) - \tfrac{1}{2}(\sigma_{xx} - \sigma_{yy}) \cos 2\theta \mp \tau_{xy} \sin 2\theta \\ \begin{pmatrix} b_x \\ b_y \end{pmatrix} &= \begin{pmatrix} \cos\theta & \sin\theta \\ -\sin\theta & \cos\theta \end{pmatrix} \begin{pmatrix} b_{\widehat{x}} \\ b_{\widehat{y}} \end{pmatrix} \end{aligned} \tag{6.38}$$

Substituting equations (6.38) into equation (6.37), we can find a new set of functions, which will become the kernels of integral equations, expressed exclusively in the local coordinate set, viz.

$$\left\{ \begin{array}{c} \tau_{\widehat{xy}} \\ \\ \sigma_{\widehat{yy}} \end{array} \right\} = \frac{\mu}{\pi(\kappa+1)} \left\{ \left\{ \begin{array}{c} K^S_{\widehat{x}}(\widehat{x}, c) \\ \\ K^N_{\widehat{x}}(\widehat{x}, c) \end{array} \right\} b_{\widehat{x}}(c) + \left\{ \begin{array}{c} K^S_{\widehat{y}}(\widehat{x}, c) \\ \\ K^N_{\widehat{y}}(\widehat{x}, c) \end{array} \right\} b_{\widehat{y}}(c) \right\}. \tag{6.39}$$

The the scheme of affices to the kernels, K, is that the subscript denotes the direction of the Burgers vector and the superscript denotes the direction of the traction (i.e. normal or shear).

We may now form two simultaneous integral equations, analogous to equations (6.23) and (6.24) for the normal crack problem, by distributing dislocations along the line of the crack. Thus

$$0 = S(\hat{x}) = \tau_T(\widehat{x}) + \frac{\mu}{\pi(\kappa+1)} \left\{ \int_0^b B_{\widehat{x}}(c) K^S_{\widehat{x}}(x, c) dc + \int_0^b B_{\widehat{y}}(c) K^S_{\widehat{y}}(x, c) dc \right\} \tag{6.40}$$

$$0 = N(\hat{x}) = \sigma_T(\widehat{x}) + \frac{\mu}{\pi(\kappa+1)} \left\{ \int_0^b B_{\widehat{x}}(c) K^N_{\widehat{x}}(x, c) dc + \int_0^b B_{\widehat{y}}(c) K^N_{\widehat{y}}(x, c) dc \right\} \tag{6.41}$$

where $\sigma_T(\widehat{x}), \tau_T(\widehat{x})$ are the values of the tractions (in $\widehat{x}, \widehat{y}$ coordinates) induced along the line of the crack by the applied loads and $B_{\widehat{x}}, B_{\widehat{y}}$ are the dislocation densities defined by

$$B_{\widehat{i}} = \frac{db_{\widehat{i}}(c)}{d\widehat{x}} \qquad i = x \text{ or } y \tag{6.42}$$

Equations (6.40) and (6.41) may now be normalised by substitutions similar to (6.25) and fundamental functions introduced analogous to (6.29), i.e.

$$B_{\widehat{i}} = \phi_{\hat{i}} \sqrt{\frac{1+r}{1-r}} \qquad i = x \text{ or } y \tag{6.43}$$

so that the discretised forms of equations (6.40) and (6.41) are

$$\frac{\mu}{\pi(\kappa+1)}\sum_{i=1}^{n}\frac{2\pi(1+r_i)}{2n+1}\left\{K^S_{\hat{x}}(s_k,r_i)\phi_{\hat{x}}(r_i)+K^S_{\hat{y}}(s_k,r_i)\phi_{\hat{y}}(r_i)\right\}=-\tau_T(s_k) \tag{6.44}$$

$$\frac{\mu}{\pi(\kappa+1)}\sum_{i=1}^{n}\frac{2\pi(1+r_i)}{2n+1}\left\{K^N_{\hat{x}}(s_k,r_i)\phi_{\hat{x}}(r_i)+K^N_{\hat{y}}(s_k,r_i)\phi_{\hat{y}}(r_i)\right\}=-\sigma_T(s_k) \tag{6.45}$$

where the integration points r_i and collocation points s_k are as given for equation (6.30). Equations (6.44) and (6.45) yield $2n$ simultaneous equations in the $2n$ unknowns $\phi_{\hat{x}}(r_i), \phi_{\hat{y}}(r_i)$. When once these values have been found the stress intensity factors may be deduced from equations similar to (6.31) i.e.

$$\begin{aligned} K_I &= 2\sqrt{2}\sqrt{\pi b}\frac{\mu}{\kappa+1}\phi_y(1) \\ K_{II} &= 2\sqrt{2}\sqrt{\pi b}\frac{\mu}{\kappa+1}\phi_x(1) \end{aligned} \tag{6.46}$$

The values of ϕ at the end points may again be found by the interpolation equation (6.32). It is perhaps worth remarking that again a check on the crack opening may be carried out using equations (6.35) and (6.36), which should, of course, be interpreted as applying in the rotated axis set, but that the corresponding shear displacement (found by replacing B_y in 6.35 by $B_{\hat{x}}$) may take any value.

Occasionally cracks may nucleate subsurface, at inclusions or other stress raisers. In these cases the crack is enclosed at both ends, and as a simplified case we will briefly consider the case of a buried crack normal to the free surface, and suffering only remote applied tension. The remote loading is assumed to be such that the crack faces are open throughout their length, and $d > b$. The general concept followed is as described earlier, but we see that in equation (6.23) the limits of the integral are changed, and hence

$$0 = N(x) = \sigma_T(x) + \frac{\mu}{\pi(\kappa+1)}\int_{d-b}^{d+b} B_y(c)K(x,c)dc \qquad d-b \leq x \leq d+b \tag{6.47}$$

The interval involved may be normalised by setting

$$\begin{aligned} r &= (c-d)/b \\ s &= (x-d)/b \end{aligned} \tag{6.48}$$

and with these substitutions the normalised form of equation (6.47) is precisely the same as equation (6.26). Now, however, a significant difference arises in the assumed form of the function $B_y(c)$. The argument embodied in equations (6.27) and (6.28), where it is shown that at the crack tip the dislocation density varies in a square root singular manner, may now be applied to *both* ends of the crack, and hence we choose

$$B_y(r) = \frac{\phi(r)}{\sqrt{1-r^2}} \tag{6.49}$$

The numerical quadrature including this required behaviour gives (see Erdogan, Gupta and Cook, 1973)

$$\frac{\mu}{\pi(\kappa+1)}\sum_{i=1}^{n}\frac{\pi}{n}K(s_k,r_i)\phi(r_i)=-\sigma_T(s_k) \tag{6.50}$$

where

$$\begin{aligned} r_i &= \cos\left(\tfrac{2i-1}{2n}\pi\right); \qquad i=1,2,...,n \\ s_k &= \cos\left(\tfrac{k}{n}\pi\right); \qquad k=1,2,...n-1 \end{aligned}$$

It will be noted that there is now one too few equations to determine $\phi(r_i)$. The last equation arises from the requirement that the crack closes at each end, i.e.

$$\Delta(b)=\int_{d-b}^{d+b}B_y(c)dc=0 \tag{6.51}$$

whose discretised form is

$$\sum_{i=1}^{n}\frac{\pi}{n}\phi(r_i)=0 \tag{6.52}$$

Lastly, the stress intensity factors may be abstracted from the value ϕ of at each end of the interval. In lieu of equation (6.31) we find

$$K_I(\pm1)=\pm2\sqrt{\pi b}\frac{\mu}{\kappa+1}\phi(\pm1) \tag{6.53}$$

where

$$\phi(j)=\frac{1}{n}\sum_{i=1}^{n}\phi(r_l)\frac{\sin\left(\frac{2n-1}{4n}(2i-1)\pi\right)}{\sin\left(\frac{2i-1}{4n}\pi\right)} \qquad j=\pm1 \tag{6.54}$$

$$\begin{aligned} l &= i && \text{if } j=1 \\ l &= n+1-i && \text{if } j=-1 \end{aligned}$$

The case of buried slant crack was also treated by Nowell and Hills (1987b).

6.3 Partially closed cracks

Cracks propelled by a combination of contact loading and some far field loading often 'see' a very complex stress history, and exist in steep stress gradients. These may be such as to press the crack faces together along at least part of their length, and for at least part of the loading cycle, which modifies the stress intensity factors experienced by the crack tip and hence influences the *range* of stress intensity controlling crack growth rate. The very simplest problem of this kind we can envisage is one of a crack normal to a free surface which is growing inwards under the action of a residual stress

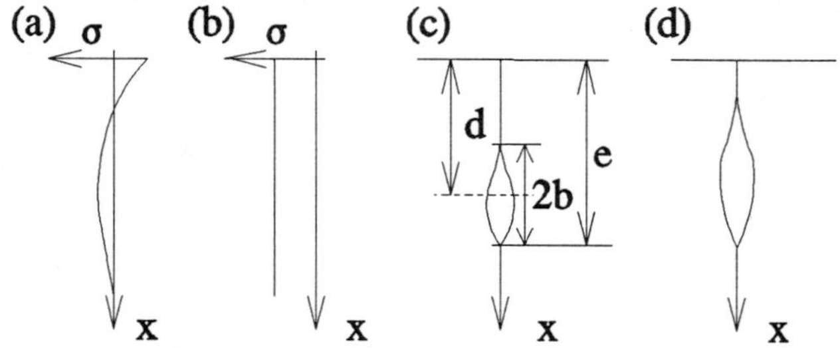

Figure 6.9: (a) Shows a typical residual stress profile left by shot peening, whilst (b) is the state of stress induced by uniform tension. (c) shows the crack shape when only stress (a) is present and (d) the crack shape when bulk tension is also applied.

field and an applied tension[3], fig. 6.9. The problem we wish to solve is, given the value of the applied stress $\sigma_T(x)$ appearing along the line of the crack in its absence, and the depth of the crack, e, what fraction of the crack length $2b$ is open, and what is the corresponding crack tip stress intensity factor?

It will be recognized that this can be achieved by starting with a formulation very similar to that used at the end of the previous section: that is, that dislocations need to be distributed over the (unknown) open portion of the crack to clear it of direct traction, and hence equation (6.47) applies, as does the normalisation (6.48) giving integral equation (6.26) which we will repeat here for convenience:

$$\frac{\mu}{\pi(\kappa+1)}\int_{-1}^{1} B_y(r)K(s,r)dr = -\sigma_T(s) \tag{6.55}$$

There is now a departure from the wholly buried crack problem however, in that we now expect that whilst the dislocation density will be singular at $x = d + b$ it will be *bounded*, indeed zero, at $x = d - b$. This is clear on physical grounds as we expect the crack faces to come smoothly together at this point, to form a cusp. The appropriate quadrature, therefore, is the one introduced in relation to the surface-breaking crack, fig. 6.8(a). Here, however, the chosen fundamental function (equation 6.29) is entirely appropriate (cf. §6.1) and the formulation is exact. The discretised form of equation (6.55) is therefore given by (6.30), with exactly the same collocation and integration points. It will be recognized that on this occasion equation (6.30) is insufficient to solve the problem, as although there are apparently n equations for the n unknowns $\phi(r_i)$ it will be recalled that the length b is unknown. The extra requirement which must be satisfied is that, at the point when the crack faces come together, equation (6.51) applies and hence its discretised form, equation (6.52),

[3]One of the advantages of the distributed dislocation technique is that it permits the uniform treatment of active and residual stresses. This point will be addresssed in more detail later in this section.

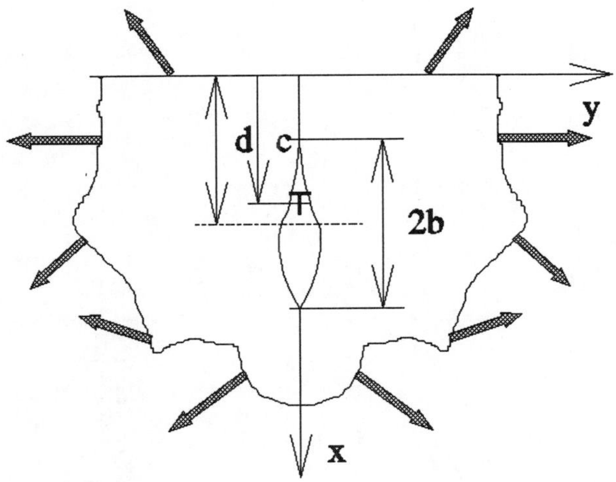

Figure 6.10: A buried crack suffering applied tension.

constitutes the additional condition required to render the problem determinate. In practice, solution is carried out by estimating the value of b (perhaps by choosing the point where the applied stress changes sign), and solving for $\phi(r_i)$ using equations (6.30). These values are then substituted into equation (6.52), and if this is not satisfied the value of b is adjusted appropriately. Convergence is usually found very rapidly.

An alternative scheme which may be employed is deliberately to guess a value of b which is too small, so that a stress intensity appears at both ends of the crack, and the geometry is effectively that shown in fig. 6.10. The analysis described by equations (6.47) to (6.54) then applies. Solutions are sought with increasing values of crack length opening, b, until the stress intensity factor at the near surface end becomes vanishingly small. Too large a value of b will result in a negative opening mode stress intensity appearing there, which has no physical meaning, but assists in the iteration. It will be found that both techniques rapidly converge to give the correct solution.

A more complex partial closure problem arises when there are shear tractions present along the plane of the crack. For illustrative purposes we shall concentrate our attention on the problem of a normal crack (Hills and Nowell, 1989) so that the effects of opening and shearing dislocations are separated, as discussed in §6.2. Figure 6.11(a) depicts a typical set of circumstances in which this kind of loading may occur. Often the shearing force Q will oscillate, thereby alternately opening and closing the crack faces. A number of possible responses may arise, as shown in fig. 6.11(b) - (e). These are;

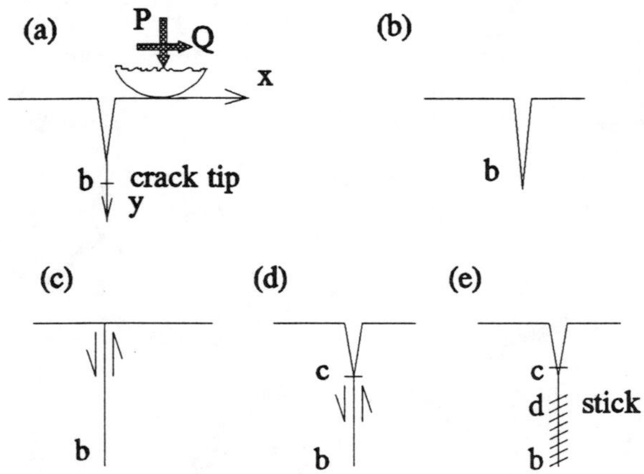

Figure 6.11: (a) Shows a normal surface breaking crack subject to both direct and shear loading. The possible crack responses are (b) open to root, (c) fully closed and sliding, (d) partially closed and sliding to root, and (e) partially closed with an adhered region.

1. Crack open to root (fig. 6.11(b)). This is the simplest possible case, already treated fully in §6.2.

2. Crack closed everywhere, but sliding (fig. 6.11(c)). This may occur if the coefficient of friction, f, between the crack faces is small. If the tractions induced by the remote loading are $\sigma_T(x)(< 0), \tau_T(x)$, slipping at every point will occur if

$$|\tau(x)| > -f\sigma_T(x) \qquad 0 \leq x \leq b \tag{6.56}$$

The unsatisfied shear traction is then given by

$$\tau_{\text{net}} = \tau_T(x) - f\sigma_T(x)sgn(\tau(x)) \tag{6.57}$$

and the integral equation to be solved for the shear-mode crack tip loading is given by (6.34), with $\tau_T(s)$ replaced by $\tau_{\text{net}}(s)$.

3. Crack partially open, but sliding to the crack tip, fig. 6.11(d). Solution here must proceed in two phases. The first step is to locate the point of closure, c, by distributing climb dislocations (b_y) in the interval $0 \leq x \leq c$. As described above, one possible technique is to guess a value of c which is deliberately too small so that a stress intensity occurs there, and to use the analysis given in §6.2 for a surface breaking crack. The value of c may be gradually increased until the stress intensity factor at $x = c$ vanishes. Alternatively a quadrature may be used which assumes bounded behaviour at each end of the interval

$(0 \le x \le c)$ and which therefore incorporates a side-condition determining c (Hills and Nowell, 1989). The second phase of the solution is to find the net shearing traction appearing along the line of the crack. This is given by

$$\begin{aligned} \tau_{\text{net}}(x) &= \tau_T(x) && 0 \le x \le c \\ &= \tau_T(x) - f\,\text{sgn}(\tau_T(x))\{\sigma_T(x) \\ &\quad + \tfrac{\mu}{\pi(\kappa+1)} \int_0^c B_y(\xi) K(x,\xi) d\xi\} && c \le x \le b \end{aligned} \tag{6.58}$$

where $K(x, \xi)$ is given by equation (6.21). The shear mode stress intensity may again be found from (6.34), providing condition (6.56) is satisfied.

4. Crack partially closed and adhered near root (fig. 6.11(e)). The first step in the solution is as given in case 3 above. The net shear is found from equation (6.58), but now it is assumed that condition (6.56) is violated near the crack tip $(d \le x \le b)$. Under these circumstances no stress intensification will result and the crack cannot grow.

Further complications arise if the crack is not perpendicular to the free surface, as both glide and climb dislocations give rise to both direct and shearing tractions. This couples the problem in quite a complex way and it is not possible to discuss all the possibilities here. An educational problem where this arises was treated by Hills and Nowell (1989). Problems of this general kind were also treated in a series of papers for the case of a crack near a contact developing uniform pressure, with oscillatory shear, (Hills and Comninou, 1985 (a, b), Sheppard, Hills and Barber (1986)). There are many other problems which have received attention using a combination of the techniques described above, and, indeed, it should be emphasized that there is no short cut to the kind of extended calculations described if reliable estimates of the stress intensity factors are to be found. It is *essential* that careful checks be performed to ensure that, where appropriate, the crack closure point(s) is properly located, and that where slipping occurs along part of the crack face not only is Amontons' friction law satisfied, but the slip direction is consistent with the sign of the frictional shear traction (Comninou, Barber and Dundurs, 1983, Chang, Comninou and Barber, 1983, Schmueser, Comninou and Dundurs 1981).

Cracks growing beneath a contact extend into a region where both direct stresses are compressive, so that there is a tendency for crack tip closure to occur. Another problem where tip closure is important is when cracks arise in components treated to produce compressive residual stress fields. Two ways of producing such a residual stress field are shot-peening and, for circular holes, split-sleeve cold expansion. Although, as has been stated, these processes produce a surface compressive layer it must be appreciated that an equilibrating tensile region must exist further beneath the surface. Thus, any crack which *does* succeed in breaking through the protective compressive region will experience an enhanced growth rate. Residual stresses have some characteristics which are different from those exhibited by ordinary 'applied' stresses, and with which the dislocation method copes well. The principal one is that as a crack grows through a residual stress field it is clear that the compressive layer is being cleaved, and hence that some relaxation must occur. However, Bueckner's theorem may be applied without modification, by imposing cancelling tractions along the line of the crack. These may, in practice, be generated by distributing dislocations. In particular, if the body is semi-infinite in extent with the same residual stress profile (e.g. fig. 6.9(a)) present at all positions, i.e. all values of y, the dislocation solution presented earlier in this subsection may be employed. No modification of the standard procedure is required, (Wilks, Nowell and Hills, 1993).

Two other sources of crack closure stresses which arise are:

(a) The effect of overloading a crack in tension. If a cracked component is loaded in tension to beyond the standard load to be applied during cyclic testing, the crack-tip plasticity zone will be correspondingly larger. When relaxation is permitted a zone of compression around the crack tip will exist and this will tend to cause closure (Ewalds and Wanhill, 1986). This problem may be avoided in testing by ensuring that the stress intensity factor increases monotonically, and that overloading is avoided.

(b) The influence of elastic mismatch on cracks propagating along an interface. This subject is far too complex for us to consider in detail here, but it can be shown (Comninou, 1977) that a crack propagating along an interface between elastically dissimilar materials will always have its tips closed, providing the boundary is physically sharp[4]. Thus, regardless of the far field loading (which might be tensile or shear in character) the crack tips will experience only a mode II singularity.

The influence of residual stresses on the fatigue behaviour of the crack is complex, and affects both the maximum stress intensity experienced and the R-ratio (figs 6.4, 6.5). Let us suppose that a crack exists in a compressive residual stress field, and is subjected to cyclic remote tensile loading, which varies between zero and some positive value. We calculate the crack tip stress intensity experienced at zero applied load, *based solely on the requirement of cancelling the tractions due to the residual stress field, and permitting crack face interpenetration.* This stress intensity factor, $K^{\min}(< 0)$ has no physical significance, as the crack faces will in reality simply be pressed together and no singularity will arise. Separately, we may calculate the stress intensity factor associated with the maximum loading *alone*, K^{LOAD}, say. Then, because the minimum value of K is zero, the true range of stress intensity experienced by the crack is

$$\Delta K = K^{\mathrm{LOAD}} + K^{\min} \tag{6.59}$$

[4]The mathematical problem which is usually solved incorporates an atomically sharp interface. The bond found between many materials arises from diffusion, so that the elastic properties vary gradually over a significant depth. In this kind of problem crack tip closure may not arise.

provided that at the point of maximum load the crack is *open at all points*. Note that this is a stronger statement than simply requiring the value of ΔK to be positive. Note also that Paris Law data are available for cracks undergoing fully reversing cycles, and for cases where the applied stress is tensile (Derby, Hills and Ruiz, 1992). For comparison with these data the R-value might be taken as

$$R = \frac{K^{\min}}{K^{\mathrm{LOAD}} + K^{\min}} \qquad (< 0) \tag{6.60}$$

The influence of R on crack propagation rate is due to a complex interaction of the applied compression with the crack tip residual stress field, and is not fully understood, but this is discussed by Ewalds and Wanhill (1986).

6.4 Three dimensional cracks

Cracks having a truly straight front are very rare, at least partly because arranging for a test geometry to be strictly two dimensional is virtually impossible, and a transition from plane stress to plane strain is inevitable. Also, it is difficult to eliminate other end effects in two dimensional problems, and, particularly for the case of contact loading, it may be preferable to design apparatus giving a well-defined three-dimensional contact pressure. In any event; cracks may nucleate at a single point and grow approximately uniformly in all directions, giving rise to a semi-circular or semi-elliptical so-called 'thumbnail' crack. It is difficult to obtain an analytical solution for geometries of this kind: exact solutions of three dimensional problems are almost confined to axi-symmetric cases, although some elliptical cracks can also be solved, and all are reviewed in the compendium by Murakami (1987). As contact loading induces complex, rapidly varying stress fields very little use may be made of standard solutions, and a numerical technique must be employed. By far the most popular method is the finite element method, whose application we will briefly discuss, although the boundary element method offers some possibilities and the three dimensional counterpart of the distributed dislocation method, the eigenstrain approach, has a great deal to recommend it; although first described ten years ago (Mura, 1982) it is developing only slowly, probably because of its apparent complexity.

A review of methods of determining stress intensity factors was given at the beginning of §6.2, and some features of the finite element method were described. It is essential, when using this technique, to use crack tip elements which show the correct asymptotic behaviour as $r \to 0$. This is normally done by moving the midside node over to the quarter point, which automatically forces the stress components to vary as $1/\sqrt{r}$. Clearly, the surrounding mesh also needs to be refined as the stress gradients will be very steep, and a combination of these two techniques will help to accelerate convergence. The stress intensity factors and J-integral may then be deduced using the processes described earlier. It is equally important to check that cracks are open at every point, and if this is not found to be true to use special 'gap' elements which

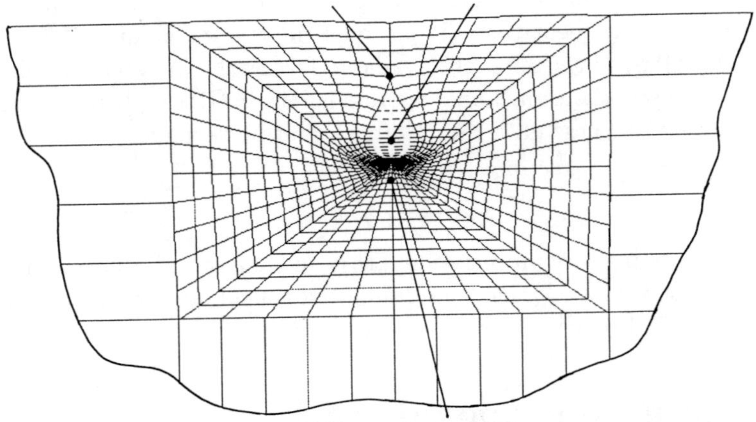

Figure 6.12: Modelling cracks using finite elements: residual stresses. Displacements are shown greatly exaggerated.

permit the contacting faces to touch, but not to overlap. Further caution should be taken in connection with the use of finite element procedures to quantify the influence of residual stresses on cracks. It is customary when modelling cracks, which are normally small in relation to other dimensions of the body, to represent the neighbourhood of the crack by a local net only, as shown in fig. 6.12. Conventionally, the applied loads would be imposed on the boundaries $y = \pm b$ as shown, and it might be thought that the residual stress distribution, e.g. as sketched in fig. 6.9(a) could simply be added in here. This is not so, however. Residual stresses, induced by plasticity, do not have to obey the requirements of compatibility but they *do* have to obey equilibrium[5]. If the residual stresses are imposed by using remote loading on the boundaries on the model, it is easy to show that they will decay towards a constant value, which is the mean applied load. If the residual stresses are self-equilibrating (i.e. have no resultant), that constant value will be zero. The residual stresses cannot, therefore, be imposed by this method, (Wilks, Nowell, and Hills, 1993), but must be installed as pre-existing stresses (with no associated strain) at each node. Most commercial finite element packages now provide this facility.

We now turn our attention to an introduction to the eigenstrain method. This is the three dimensional counterpart of the dislocation density method, and is still in the early stages of development, with only a handful of papers published in the research literature, most of them for the purposes of calibrating the technique. The three-dimensional analogue of the dislocation is the eigenstrain itself ϵ^*_{mn} which varies

[5]Compatibility of displacements is, of course, required but the plastic deformation involved in the introduction of residual stresses means that it is not possible to use Hooke's law to formulate the compatibility equation in terms of stresses (as is done in elasticity theory).

from point to point and exists over the surface of the crack or indeed throughout the volume of an inclusion. In order to find its influence on the state of stress within a body we first write down the Green's function, G, (which is known in closed form for an infinite space or half-space), for the displacement, u_j, arising at point (x_1, x_2, x_3) due to a force $F_i(y_1, y_2, y_3)$ i.e.

$$u_j(x_1, x_2, x_3) = G_{ij}(x_1, x_2, x_3, y_1, y_2, y_3)F_i(y_1, y_2, y_3) \tag{6.61}$$

where i, j denote the directions of the force and displacements respectively and, in Cartesian co-ordinates, we identify the subscripts 1, 2, 3 with x, y, z respectively. Next, we introduce the elasticity matrix, C_{ijkl}, which relates the stresses and strains, i.e.

$$\sigma_{ij} = C_{ijkl}\epsilon_{kl} \tag{6.62}$$

For an isotropic material the non-zero elasticity elements of the elasticity matrix are found from the more familiar elastic constants by noting that

$$\begin{aligned} \sigma_{ii} &= 2\mu\epsilon_{ii} + \lambda(\epsilon_{11} + \epsilon_{22} + \epsilon_{33}) \qquad & i = 1, 2, 3 \\ \sigma_{ij} &= 2\mu\epsilon_{ij} & i \neq j \end{aligned} \tag{6.63}$$

where μ is the modulus of rigidity and λ, Lamé's constant, is given by

$$\lambda = \frac{2\mu\nu}{1 - 2\nu} \tag{6.64}$$

The most general form of the eigenstrain compatibility principle is found by differentiating ϵ_{ij} with respect to the lth component of direction at the point where the eigenstrain $\epsilon^*_{mn}(y_1, y_2, y_3)$ is to be installed, so that, substituting (6.62) into (6.61) and integrating over the volume, V, occupied by the eigenstrains we get

$$u_j(x_1, x_2, x_3) = \int_V C_{ilmn} \frac{\partial G_{ij}}{\partial y_l} \epsilon^*_{mn}(y_1, y_2, y_3) dV \tag{6.65}$$

Usually, we need to know the state of stress induced rather than the displacement field. This is readily found using the strain displacement relation and Hooke's Law (equation 6.62), so that

$$\sigma_{jk}(x_1, x_2, x_3) = \int_V C_{ilmn} \frac{\partial T_{jki}}{\partial y_l} \epsilon^*_{mn}(y_1, y_2, y_3) dV \tag{6.66}$$

where

$$T_{jki} = C_{jkmn} \frac{\partial G_{im}}{\partial x_n} \tag{6.67}$$

We now wish to collapse the volume of the strained region in the x_3 direction to $\pm h/2$ so that the volume reduces to a prism, and then take the limit $h \to 0$ so that the prism reduces to a crack. With this limit, equation (6.66) reduces to

$$\sigma_{jk}(x_1, x_2) = \int_S C_{ilm3} \frac{\partial T_{jki}}{\partial y_l} b_m(y_1, y_2) dA \tag{6.68}$$

where we have also now chosen the field point to lie in the x_3 plane, and the three quantities b_m to represent the displacement discontinuities across the plane of the crack. The opening mode is

$$b_3 = \lim_{h\to 0} \int_{-h/2}^{h/2} \epsilon_{33}^* dy_3 \tag{6.69}$$

and the two shearing modes are

$$b_\alpha = 2 \lim_{h\to 0} \int_{-h/2}^{h/2} \epsilon_{\alpha 3}^* dy_3 \qquad \alpha = 1,2 \tag{6.70}$$

Applying Bueckner's theorem, if the tractions occurring in the plane of the crack in its absence are σ_{3n}^0, we see that the integral equation requiring the crack faces to be traction-free is

$$-\sigma_{3n}^0(x_1, x_2) = \int_S K_{nm}(x_1, x_2, y_1, y_2) b_m(y_1, y_2) dA \tag{6.71}$$

where the kernels required for use in equation (6.71) are

$$K_{nm} = C_{ilm3} \frac{\partial T_{3ni}}{\partial y_l}(x_1, x_2, y_1, y_2) \tag{6.72}$$

Values of T_{jki} for inclusion in (6.72) are available for an infinite space, the so called Kelvin solution (Timoshenko and Goodier, 1951), and for a half-space (Mindlin, 1936). Implementation of equation (6.71) is considerably more complex than the two dimensional version. First, the kernel now has not a Cauchy singularity ($1/r$) but a much stronger ($1/r^3$) characteristic, which is very difficult to handle efficiently numerically. Secondly, the domain of integration, S, i.e. the area delineating the crack, must be divided into finite sized elements for a numerical quadrature to be implemented. Lastly a weight function has to be introduced which gives the correct asymptotic form for the opening (or shear displacement) as the crack front is approached. This is done by writing (Murakami and Nemat Nasser, 1983)

$$b_n(y_1, y_2) = W(y_1, y_2) f_n \tag{6.73}$$

where

$$W = \sqrt{2b\epsilon(y_1, y_2) - \epsilon^2(y_1, y_2)} \tag{6.74}$$

b is a representative crack length and ϵ is the closest distance from the integration point to the crack front. The technique is fully described by Mura (1982) and an improved integration technique for evaluating the kernel is introduced by Dai, Nowell, and Hills (1993).

Chapter 7

Fretting Fatigue Tests

7.1 Introduction

So far, we have presented several of the elements of calculation necessary to analyse a range of idealised fretting contacts. They merely provide tools for determining the state of stress, strain and displacement, but do not, in themselves, permit production of initiation or crack propagation criteria, without extensive experimental data. The object of fretting fatigue tests must be to permit monitoring of fretting fatigue crack propagation and, more importantly, initiation, under laboratory conditions, so that the influence of different load histories, surface treatments, surface finishes and indeed underlying materials may be found. There are three general categories of tests which might be envisaged, viz.

1. Simulation of a real engineering fretting problem. In tests of this kind the intention is to reproduce as carefully as possible, either full size or possibly on a reduced scale, the contact problem arising in the prototype. This clearly has a big advantage over other kinds of tests insofar as there is a real attempt to simulate the conditions of the fretting contact themselves as accurately as possible, and so results found from the test apparatus could be carried over without modification to the real engineering problem.

2. Materials ranking tests. The intention of these tests is to assess a material's resistance to fretting fatigue in some general way. A simplified geometry is chosen so that accelerated tests may be conducted, but little attention is paid to the details of the contact conditions. Instead, it is assumed that the same contact geometry is repeated for every test. It is also tacitly assumed that the materials ranking may be carried over to other kinds of fretting contact. The advantage of this kind of experiment is that the tests may be completed very rapidly.

3. Idealised fretting fatigue tests. The basic philosophy here is to carry out a series of tests under very well defined, simple contact geometries. These are chosen so

that it is possible to deduce in closed form as many of the pertinent variables (e.g. stress, slip, displacement) as possible. The object is to understand as far as possible the mechanics of the fretting fatigue process. Tests in this category are therefore quite different from others cited above, as the results are intended to be applicable to a wide range of geometries quite different from those used in the tests themselves.

Each category of tests has its merits, and serves a different purpose. It might be remarked in connection with tests to simulate real engineering contacts that a potential difficulty arises if the *size* of the contact used in the simulation is very different from the prototype. This is because two length scales other than the dimension of the contact itself enter the problem, viz. the amplitude of displacement of the contact and, less obviously, the roughness or asperity scale. In particular, it should be noted that the characteristic asperity pitch influences the number of load cycles experienced on an asperity scale. This point is discussed in more detail in §9.4, but will be summarised here: if the asperities are assumed to be regularly spaced, and the amplitude of tangential relative displacement equal to, say, three asperity spacings, each surface particle in the contacting body will experience three reversals of asperity load (i.e. cycles), per overall load cycle. Thus, if an experiment is conducted using material with the same surface finish as the prototype, but with a contact dimension which is much smaller, it is possible that the number of load reversals experienced on an asperity scale is *lower* than the prototype, because of this effect.

Further, it may be very difficult to reproduce exactly the real contact, particularly if it is complete, because of the profound influence of minor surface damage (§2.1). It is very probable that, in any batch of nominally identical components, the spread of surface texture and minor defects associated with machining tolerance, together the natural occurrence of burrs, would mean that the true contact conditions would vary widely from contact-pair to contact-pair, when these components were employed in an experiment. For example, in a flanged joint, the presence of some small surface 'bump' (perhaps a piece of trapped swarf) would cause separation locally, and an increased contact pressure elsewhere. Further, beneath the bump itself the contact pressure would reach a very high value, and probably cause local yielding. Effects of this kind, which perturb the contact pressure greatly from the value predicted by an idealised analysis, give rise to considerable experimental scatter. These points are discussed in more detail in §9.4.

7.2 Bridge-type tests

In principle any fretting test may be carried out either in the sliding or partial slip regimes. In practice it is very difficult to devise a well-controlled test where the magnitude of the imposed displacement is determined solely by an external actuator or cam. This is at least partly because the compliance of any linkage used will often be at

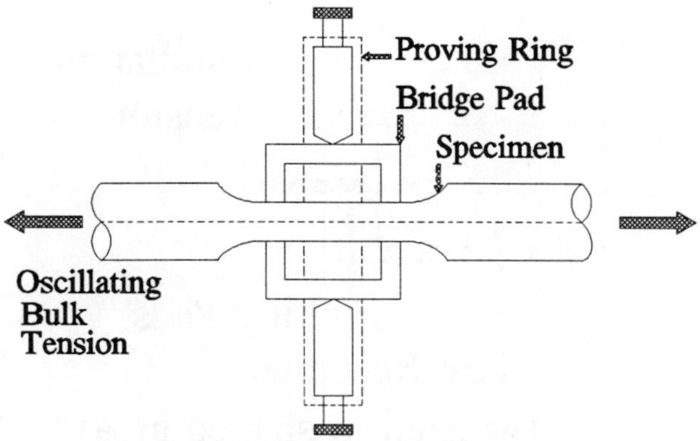

Figure 7.1: Bridge and tension-test apparatus.

least as great as the compliance of the contact, and also because it is extremely difficult to measure displacements reliably down to the level of about $10\mu m$. We therefore consider it preferable to design tests which exploit the compliance within the specimen pair itself to introduce the necessary tangential displacement, as the bulk loading is applied. There are two approaches to this: one is to use a convex (incomplete) contact operating in the partial slip regime, so that there is a central adhered portion where no relative slip occurs. Surrounding this is a slip region (chapter 3) where the relative slip amplitude may be reliably found from a calculation of the deformation in the contacting bodies. The second approach is to use a specimen with two physically quite separate contacts connected in the form of a 'bridge', and to use the compliance of either the bridge or the contacting tensile test specimen as a means of introducing relative displacement. It is this class of experiment with which we are concerned here. Almost invariably the bridges have contacting pads which have flat faces, and therefore form complete contacts.

The 'bridge' method of inducing slip is illustrated schematically in fig. 7.1. The two bridge specimens are clamped, using a proving ring, onto a simple tensile test specimen, which is then loaded cyclically in a hydraulic testing machine. The specimens should have sufficient axial cross-section to withstand axial compression, although there is no need for the major applied load to be fully reversing. A proving ring is commonly used to apply the normal contact force, and if this is relatively compliant in comparison with the bridge stiffness there will be little drop off in load should any wear occur. Points on the tensile test specimen, beneath the feet of the bridges, will separate because of the axial bulk strain induced, thereby imposing a shearing traction on the contact pads. The amplitude of relative displacement is controlled by extension of the tensile specimen, less any stretching which occurs in the bridges. In the standard arrangement sketched, the bridge-arrangement is left floating, and

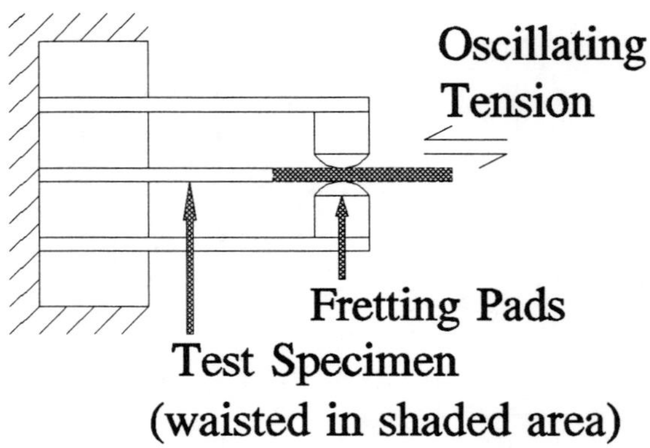

Figure 7.2: Fretting apparatus using cantilever springs to apply the normal contact force. An oscillating bulk tension is also applied.

hence is free to displace axially as a rigid body by an arbitrary amount. In a perfect system, all four contacting pads would exhibit the same finish, and cleanliness and hence the same coefficient of friction, and they would therefore exert the same shearing force and experience the same slip. In practice this may not be achieved, and one pair of feet may adhere whilst the other pair accommodates the whole of the imposed displacement. This problem may be averted by using a bridge-type configuration in which one end of the bridge is rigidly fixed, and the other, carrying the fretting pads, bears on the specimen (fig. 7.2). Also, in the geometry sketched, the pads are shown as cylinders, and this overcomes the problem of sensitivity to machining imperfections and to localized (but difficult to quantify) plasticity alluded to earlier. As an alternative to applying the bulk load in simple tension, a transverse force is sometimes applied, as shown in fig. 7.3. This produces a different stick-slip regime, and contact pressure distribution, but again it is difficult to quantify.. It is easy to see intuitively that a singularity in the contact pressure must occur at the edge of the contact.

All of the above tests rely on a reciprocating load to apply the bulk loading to the specimen. This may be done readily using a servo-hydraulic test machine, but inertia of the moving parts will limit the cyclic speed which may be attained, and higher speeds will also mean higher inertia forces with attendant uncertainty in the true load imposed on the working length of the specimen. An ingenious way of removing the need for reciprocating loads is to use a four-point rotating bending specimen, of the type pioneered over a century ago by Wöhler, and clamping bridge specimens to the gauge length, fig. 7.4. As the rotating specimen turns, elements in the surface alternately experience tension and compression, and there is therefore a corresponding elongation or compression. Just as in the configurations described above involving oscillating tension, there is a tendency for the feet of the bridge to open and close

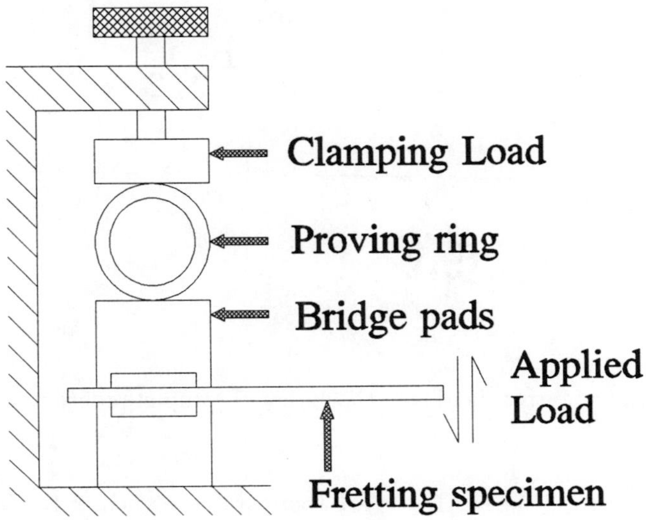

Figure 7.3: Fretting apparatus in which the bulk state of stress is induced by the application of a cyclically varying transverse bending load.

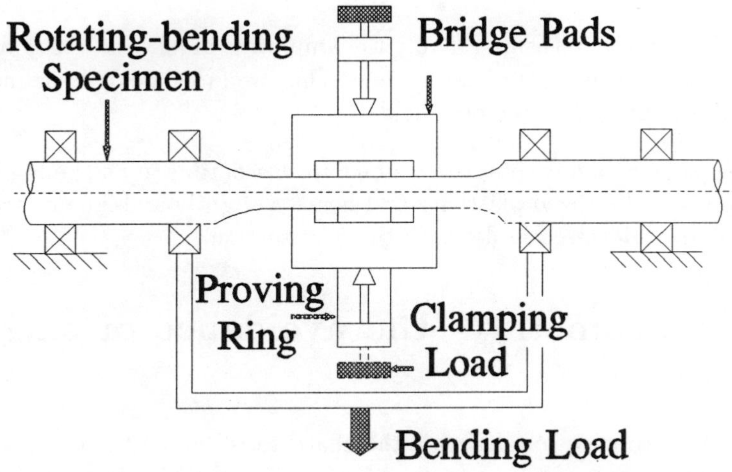

Figure 7.4: Rotating-bending type test apparatus.

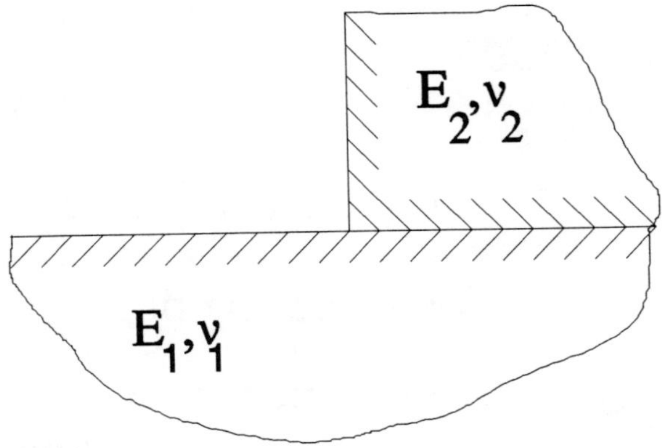

Figure 7.5: Asymptotic stress field near the corner of complete contact: nomenclature.

as the specimen rotates. Shear stresses are therefore developed at the interface. The principal difficulty here (aside from considerations of contact pressure distribution) is balancing the bridge/proving ring assembly, which is essential to avoid unwanted inertia forces. However, if this can be achieved, it does provide a means of imposing a very high rate of cyclic loading with little external expenditure of energy. The two fundamental points which emerge from this discussion may be reiterated in the following way:

1. The easiest way to measure reliably the amplitude of particle slip is to operate the apparatus in the partial slip regime. This produces a closely defined stress, strain and relative displacement field.

2. Fretting pads involving complete contact are susceptible to machining imperfections, produce localized plasticity and hence a slightly ambiguous stress state. This point is addressed in detail in the next section.

7.3 Test geometry - the avoidance of singularities

The first problem we wish to address is the characterisation of the contact problem beneath a square-ended pad, one corner of which is shown in detail in fig. 7.5. It was shown in Chapter 2 that a solution to the contact problem as a whole cannot be obtained in anything like a closed form if both pad and contacting half-plane have a finite stiffness, and the only solution to date of this problem was provided by Khadem and O'Connor (1969), for frictionless and adhesive cases. However, the use of asymptotic analysis does enable us to characterise the stress state near the corners of the pad, and this gives useful physical insight into the nature of the contact tractions.

The first analysis of this class of problem was by Bogy (1968, 1971) who looked at the stress state near the apex of two bonded wedges, which in general may be elastically dissimilar. He showed that there are four possible classes of stress state at the tip of the wedge, viz.

(a) The state of stress may be bounded and of order one.

(b) It may be singular and of order $\ln(r)$.

(c) It may be singular and oscillatory, varying like $r^{\lambda-1}\cos(\ln r)$ or $r^{\lambda-1}\sin(\ln r)$.

(d) It may be singular but not oscillatory, varying like $r^{\lambda-1}$.

Which of these four cases obtains depends on the root of a determinant, which is a function of the two wedge angles and the two Dundurs' constants α, β. These are defined by[1]

$$\alpha = \frac{\Gamma\left(\kappa_1+1\right)-\left(\kappa_2+1\right)}{\Gamma\left(\kappa_1+1\right)+\left(\kappa_2+1\right)} \tag{7.1}$$

$$\beta = \frac{\Gamma\left(\kappa_1-1\right)-\left(\kappa_2-1\right)}{\Gamma\left(\kappa_1+1\right)+\left(\kappa_2+1\right)} \tag{7.2}$$

where

$$\Gamma = \mu_2/\mu_1$$

and

$$\begin{aligned} \kappa &= 3-4\nu && \text{in plane strain} \\ &= \left(3-\nu\right)/\left(1+\nu\right) && \text{in plane stress} \end{aligned}$$

For the case when one wedge angle is π and the other is $\pi/2$, as they are in fig. 7.5, there is always a singularity at the apex and case (d) above obtains, with $\lambda < 1$. For a complete range of values of $\lambda(\alpha, \beta)$ Bogy's papers may be consulted, but if the two bodies are elastically similar so that $\alpha = \beta = 0$, $\lambda = 0.55$. If the pad is more rigid than the half-plane the singularity is stronger, and in particular if it is rigid $\lambda = 0.5$. Conversely, if the half plane is more rigid than the pad the singularity is weaker; if the half-plane is incompressible $\lambda = 0.71$, which is its largest possible value for this geometry.

The opposite limit of the contact under *frictionless* conditions was studied by Dundurs and Lee (1972). In this case there is a further reduced dependence on the elastic constants; the solution depends only on α. It is shown that if the pad and half-plane are elastically similar, $\lambda = 0.77$. If the pad is rigid, the classical Boussinesq

[1]Only one constant, β, arises in the formulation of two-dimensional or axi-symmetric contact problems (see §2.2, equation (2.18)). For a further discussion of both constants refer to Barber (1992).

solution (Chapter 2) applies and $\lambda = 0.5$. However, if the pad is elastic but the half-plane rigid ($\alpha = -1$) case (a) applies and the stress state is finite. Note that this is the only geometry which gives rise to a bounded state of stress, and that if the half-plane has a finite stiffness the singularity re-appears. If the contact between the pad and the half-plane sustains a finite *frictional* effect, analysis by Comninou (1976) applies. The results of this work show that normally a singularity of type (d) above arises, but that if the pad is rigid the state of stress oscillates in the manner described above as case (c).

The lesson to be learned from the above analysis is that a singularity is *inevitable* if a flat-ended pad is used. The exact details of the nature of the singularity will change, depending on the combination of elastic constants, the coefficient of friction and indeed the included angle at the edge of the pad if this is not $\pi/2$. In practice the singularity would be relieved by plasticity. If misalignment occurred and the pad slightly tilted, so that the contacting faces separated, the contact pressure would fall continuously to zero there, as the geometry would be 'incomplete'. These, then, are the only two possible end conditions at the edges of a contact, save for the very special, and somewhat artificial case of an elastic block pressed onto a rigid half-plane. These results were also revealed by a careful boundary element and experimental investigation by Sato (1992).

The use of convex fretting pads, giving rise to incomplete contacts, averts the above difficulties, as the pressure distribution falls continuously to zero at each edge of the contact. Therefore, providing that the load is within the elastic limit (Hills, Nowell and Sackfield, 1993) the contact will be everywhere elastic and the resulting stress field carefully defined. Also, although manufacturing errors (burrs, swarf, nicks) will have a local influence they will not alter the overall pressure distribution unless they are extremely severe.

If a convex fretting pad is to be used, the next decision to be made is whether it should be axi-symmetric or two dimensional. If the latter is chosen it has the merit of permitting a very ready elastic analysis in closed form, which is probably exact. No approximations whatever need be made, and the influence of elastic mismatch may be incorporated (see §4.2). Further, the slip direction of surface particles will lie in the plane of the contact, and there can be no transverse component of displacement. Yet another advantage is that if the test specimen is made relatively narrow, cracks whose fronts are approximately straight may be anticipated, fig. 7.6(a) permitting easy and convenient monitoring of the crack size using optical methods.

Against the advantages cited above, must be the inevitable problems of the edge-effects which are always present when two-dimensional configurations are realised: no contact is infinitely long, and hence there will always be a three dimensional stress state present. The best that can be done is to make the contact pads and transverse width of the specimen the same, as if either is made wider than the other there will certainly be severe stress concentration effect at the ends of the contact. If the two are made the same width, fig. 7.6(a), the remaining problem is that of the condition

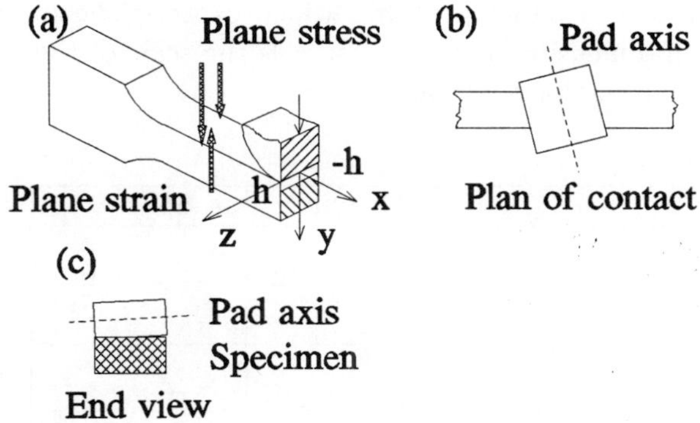

Figure 7.6: Plane incomplete contact: (a) plane stress to plane strain transition, (b) skew misaligned pad, (c) tilted misaligned pad.

in the z-direction. On the faces $z = \pm h$ plane stress clearly applies, as $\sigma_{zz} = 0$, but as we move towards the central plane the state of stress will move more nearly to one of plane strain. True plane strain would only apply as the ratio $h/2a$ (where a is the contact half-width) became large, and this is at variance with the suggestion made above that narrow specimens allow cracks with approximately straight fronts to be grown. Nevertheless, Johnson(1985) comments that the contact pressure under plane stress conditions will be less than under plane strain, and that therefore the severest loading will be remote from the ends. Another potentially serious problem with using a two dimensional configuration is that of misalignment. This can take two forms, sketched in fig. 7.6(b, c), viz. the pad can be pulled skew to the axis of the specimen, or the normal load may be offset, giving rise to a contact strip which tapers across the width of the specimen. The former induces edge stress concentrations, promoting a corner-initiated failure.

The use of an axi-symmetric fretting pad removes all of the problems of misalignment and edge effects, but at the expense of introducing a much more complex contact problem and also a small transverse component of displacement in the slip zone. If a spherical contact is used (Kuno et al., 1989) the severest state of stress lies on the plane of symmetry, and is, at least, well-defined. On the central plane the slip direction is always aligned with the direction of the applied shearing force. Cracks therefore tend to initiate here and to grow into an approximately thumbnail-shape. They are therefore very difficult to monitor, unless a metallurgical section is taken at the end of the test, and measuring the growth rate is extremely difficult.

It is also useful to consider the various scaling effects when deciding whether to chose an axi-symmetric or plane geometry. In what follows the usual symbols will be used: a is the radius of contact or contact semi-width, R is the relative radius of curvature, A is the composite compliance (equation 2.17), and p_0 the peak contact pressure. The two important normalizing parameters which govern the environment at the fretting site are the contact pressure, which controls the magnitude of the stresses and the length dimension, a, which controls the scale of the stress field and hence the volume of material loaded. These are related by

	Cylinders	Spheres
Size	$a^2 = \frac{2PAR}{\pi}$	$a^3 = \frac{3PAR}{8}$
Peak contact pressure	$p_0 = \sqrt{\frac{2P}{\pi AR}}$	$p_0 = \frac{2}{\pi}\sqrt[3]{\frac{3P}{A^2R^2}}$

Investigations by Bramhall (1973) and Nowell (see Nowell and Hills, 1990) looked for a size effect in fretting fatigue. It was therefore necessary to keep p_0 constant, so that the magnitude of the stress state remained constant, but to vary the contact size, by adjusting the characteristic contact dimension. As may be seen from the above table, in the case of line contact of cylinders these requirements mean that P/R must be kept constant whereas for point contact of spheres P/R^2 must be maintained constant. The latter would require an impossibly wide range of load. Interestingly, for a 'linear' indenter (cone or wedge) of external angle ϕ (see §2.3) the corresponding results are

	Wedges	Cones
Size	$a = \frac{PA}{2\phi}$	$a^2 = \frac{PA}{2\pi\phi}$
Mean contact pressure	$\bar{p} = \frac{\phi}{A}$	$\bar{p} = \frac{2\phi}{A}$

To the authors' knowledge this geometry has never been exploited in a fretting fatigue test, although it would seem to have some potential.

7.4 Fretting tests based on the Hertzian contact

For the reasons set out in the previous section, the Hertzian contact geometry has much to commend it, albeit at the expense of experimental complication. A number of tests have been conducted at Oxford based on this configuration, and which have

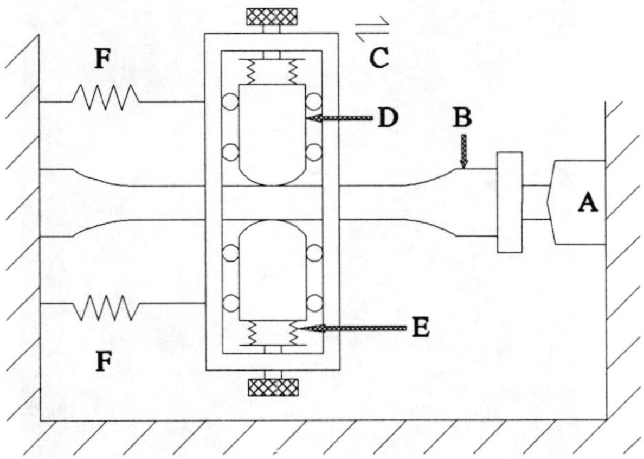

Figure 7.7: Fretting fatigue apparatus with controlled partial slip using a single axis hydraulic testing machine.

permitted an elucidation of the influence of the physical contact conditions on fretting fatigue in ever finer detail.

The first two versions of the apparatus may be represented schematically by the layout shown in fig. 7.7, and were designed to be used on a single-actuator hydraulic test machine. Actuator A applies a cyclically varying load to specimens B, and under normal conditions the force developed is fully reversing. Part way along the specimen a cradle C carries a pair of fretting pads D which are clamped onto the specimen by compliant springs E, whose compression may be adjusted to give the required normal contact load. As the specimen is cycled, there is a tendency for the cradle and pads to move back and forth as a result of the strains developed in the tensile test specimen to the left of the cradle. This motion is inhibited by springs F which therefore develop a force proportional to the displacement of the cradle, and is manifested as a shearing force on the fretting pads. As the load in the main specimen is fully reversing, the shear force history also has this characteristic, and is in perfect synchronism. In practice it is difficult to make the combined stiffness of the springs F and cradle sufficiently high to develop a large shearing force at low levels of bulk tension, and the apparatus is therefore principally used at high nominal stress levels. It will be recognised that, to maintain equilibrium, the sum of the tension force and shearing force to the left of the cradle must balance the tension force alone to the right. In practice however, the cross sectional area of the main specimen will be much larger than the contact area, so that although the bulk tension and contact shearing stresses will be of the same order, the frictional shear force will be relatively low, and it is a satisfactory approximation in the analysis to assume that the tension in the main specimen is constant along its gauge length.

Figure 7.8: Realisation of the fretting apparatus shown schematically in fig.7.7.

Two realisations of the apparatus were devised by Bramhall (1973), and by Nowell (1988), (see fig. 7.8). The latter tests were carried out under more carefully controlled conditions, but were motivated by those of the former, in which an apparent size effect was discovered, and which was confirmed. The need to test for a size effect meant that the choice of indenter was restricted to a two dimensional geometry (see §7.3) in order to ensure that the range of P/R used was physically acceptable. Had it not been necessary to maintain p_0 constant, it would clearly have been possible to use axi-symmetric or barrelled indenters in order to minimise edge effects. Pressure sensitive paper was used to adjust the alignment of each pad, to ensure that the contact in every case was as even as possible. The only variation needed on a classical Mindlin-Cattaneo partial slip analysis was to incorporate the effect of the cyclically varying tension, and this is described in §3.5.

It has been mentioned that the first two versions of this fretting apparatus were used to investigate fretting fatigue at high bulk stress levels. This is an unavoidable consequence of inducing the shearing force by means of springs, as the displacement

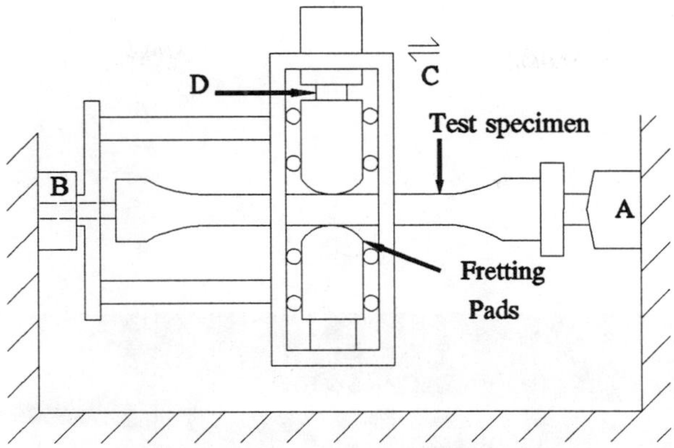

Figure 7.9: Fretting fatigue apparatus using a two-actuator hydraulic testing machine.

of the pads along the line of the bulk loading is inevitably very small, and there is a limit to the stiffness which can be designed into the beams constituting springs F (figs 7.7, 7.8). A further limitation of the apparatus is that the shearing force is always developed in phase with the bulk loading. It is therefore not possible to investigate the effect of phase differences or frequency differences, which bring with them the possibility of explaining initiation sites on either side of the fretting pads, or varying the number of asperity passes with independent control of the bulk tension. Also, it is not possible with the simpler apparatus to keep the bulk tension constant and to cycle only the shear force.

All of these problems may be overcome by using a test machine having two actuators, as shown in fig. 7.9. In this apparatus, whose physical appearance is shown in fig. 7.10, the actuator A controls the bulk tension in the specimen alone. A second actuator, B, is connected to a cradle, C, carrying the fretting pad arrangement. The two pads may float normal to the specimen, but are pressed together by a small piston, D, which is loaded by means of a hand pump with an accumulator. This arrangement permits all of the possibilities cited above to be realised. In particular, by running the bulk tension actuator at double the frequency of the shearing force, we may generate a tensile environment at each edge of the contact patch in phase with the shearing forces. This gives twice the number of possible initiations sites, but modifies the stick slip regime experienced by the contact.

In conclusion, the various kinds of fretting apparatus used fall into two general categories. Those which use a rotating-bending configuration are very useful as materials-ranking tests, because the cyclic rate may be made very high without incurring any inertia forces. However, they normally have non-ideal, flat-ended fretting pads which develop poorly defined contact pressure distributions. The second kind of test uses

Figure 7.10: Realisation of the two-actuator fretting fatigue experiment shown schematically in fig. 7.9.

some kind of reciprocating motion to imposed the fretting force. This may either be independently controlled or induced by deformation of the principal test specimen. In either case the apparatus is normally run under partial slip conditions, so that the relative slip of surface particles is determined by the geometry and elasticity of the contacting bodies. In apparatus run in the sliding mode, independent control of the surface displacement is, in principle, possible, but may be difficult to achieve at the magnitudes needed.

Chapter 8

Analysis of crack propagation

8.1 Introduction

Fatigue cracks may propagate either from an existing flaw, such as an inclusion or a defect in a weld, or they may grow from damage nucleated as part of the early stages of fatigue itself. It is into this second category that *fretting* fatigue falls. The fretting process must first initiate an embryo crack and then propagate it in order to produce a failure. The life of a component suffering fretting fatigue may therefore be conveniently divided into initiation and propagation phases. In contrast to plain fatigue the initiation phase of fretting fatigue life is often, although not always, quite short. In reality the term fretting fatigue encompasses a range of conditions from mild contact tractions accompanying bulk stress amplitudes sufficient to cause failure in plain fatigue, to severe fretting in the presence of relatively low bulk stresses. The spectrum of loading conditions can be conveniently summarised in a diagram such as that shown in fig. 8.1. It is possible for fretting fatigue to affect significantly *both* crack initiation and crack propagation. Thus, a full understanding of fretting fatigue requires a consideration of the effect of fretting on both initiation and propagation phases of crack life. Initiation occurs at a microscopic scale and a detailed understanding can be achieved only by a micromechanics analysis, although some progress can be made by considering bulk properties of the contact. In contrast, once a crack has developed and is larger than several material grain sizes, it should be possible to explain its propagation by employing the same techniques of fracture mechanics as are used for other types of fatigue. Indeed, crack growth depends entirely on conditions at the crack tip and it is impossible for a crack to 'distinguish' whether the propelling stresses arise from a contact loading or from some other far field. Thus, the analysis of crack propagation would appear to be far more tractable than that of initiation and it will be addressed first. Crack initiation will be discussed further in Chapter 9.

The objective of any analysis of crack propagation is to predict the direction and rate of crack growth under some known applied loading. Much work, both

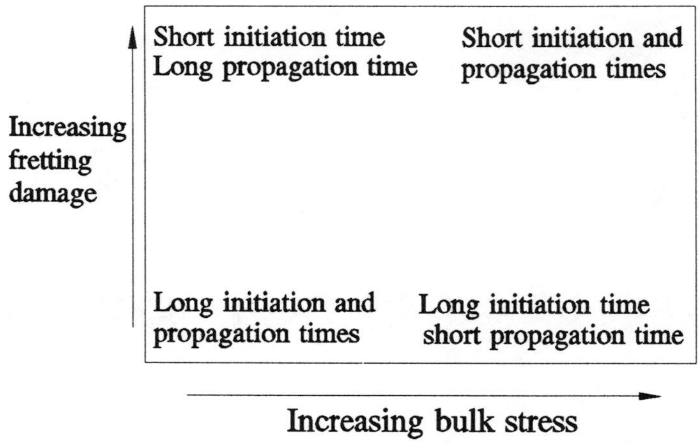

Figure 8.1: The spectrum of loading conditions encompassed by the term fretting fatigue.

experimental and analytical, has been carried out over the last thirty years in this area and for simple loading conditions it is possible to predict both the direction and rate of growth with some confidence. Unfortunately the situation is less clear for complex mixed-mode loading as encountered in fretting fatigue, and it is therefore appropriate to begin by reviewing the state of understanding for simpler cases before turning to the application of this knowledge to the fretting problem. As stated in Chapter 6, the life of a crack can be conveniently divided into a number of stages, shown in fig. 8.2. In Stage 1 (using a nomenclature first suggested by Forsyth (1961)) crack growth takes place by shear deformation in localised slip bands near the crack tip (Suresh (1991)) and the process is primarily controlled by shear. Once the crack has reached a few grain diameters in size, transition to Stage 2 behaviour takes place. Here the crack grows primarily as a result of mode I loading[1] at the crack tip, and when plotted on logarithmic scales a linear relationship between growth rate and applied stress intensity factor range ΔK is revealed. Eventually the crack reaches a third stage of its life when the high ΔK values cause acceleration in the crack growth rate towards failure. This regime has been termed Stage 3 in fig. 8.2, although this was not part of Forsyth's original nomenclature.

It is Stage 2 behaviour which often dominates crack life and it is in this regime that fracture mechanics is most useful in predicting crack growth rates. Although crack growth will, in reality, take place by plastic deformation in a 'process zone' at the crack tip, the characteristics of this zone will be controlled by the surrounding elastic field, and thus the elastic stress intensity factors are likely to prove useful in predicting crack behaviour. As mentioned in Chapter 6, experimental studies show

[1]See Chapter 6, page 126

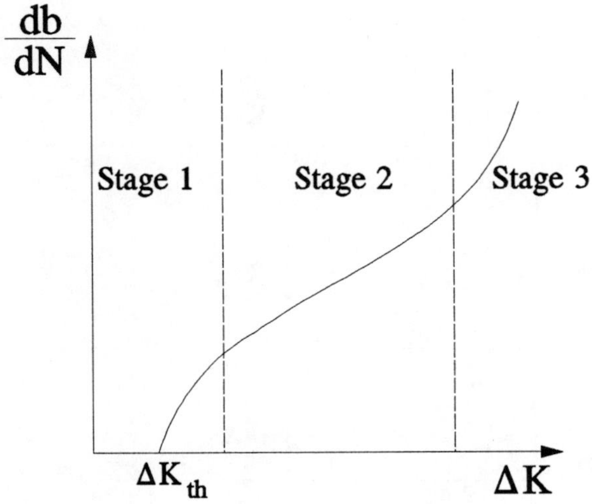

Figure 8.2: Plot of crack growth rate $\frac{db}{dN}$ against stress intensity factor range ΔK, showing the three stages of crack growth.

that it is not the maximum stress intensity factor itself which is the most useful parameter in quantifying cyclic loading, but the stress intensity factor *range*, ΔK, defined by

$$\Delta K = K_{\max} - K_{\min} \tag{8.1}$$

Where $K_{\max}$ and $K_{\min}$ represent the maximum and minimum values of K experienced during the loading cycle. Since, in Stage 2, cracks grow largely in Mode I, these quantities are normally taken as the Mode I stress intensity factors. It has already been noted that, in Stage 2, the relationship between crack growth rate, $\frac{db}{dN}$ and ΔK can be adequately described for many materials by the Paris Law:

$$\frac{db}{dN} = A(\Delta K)^m \tag{8.2}$$

The two constants A and m vary with both material and environment, although tabulated values can often be found in the literature (e.g. Derby, Hills, and Ruiz (1992)). Unfortunately, however, they are also found to vary with the *mean* stress intensity factor as well as the range. Consider, for example, a crack growing under a cyclic stress σ which is uniform (with respect to position) remote from the crack. The two loadings shown in fig. 8.3(a) and (b) have the same stress intensity factor range ΔK, but are found experimentally to produce different growth rates. Rather than relate these differences to the mean value of K it is customary to use an alternative parameter, the R-ratio, defined by

$$R = \frac{K_{\min}}{K_{\max}} \tag{8.3}$$

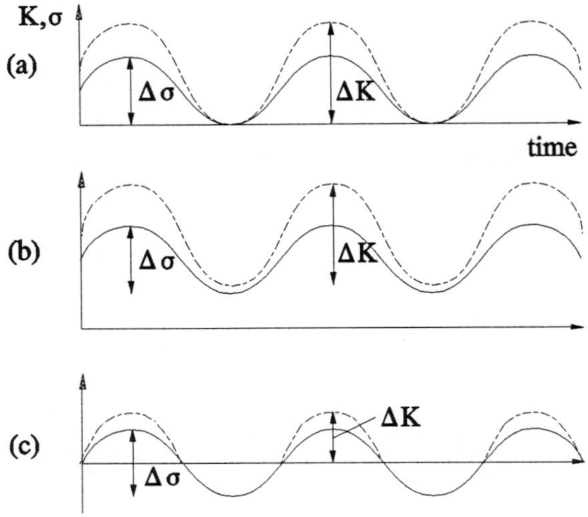

Figure 8.3: Different cyclic loading histories (a) stress varying from zero to $\sigma_{\max}$ (b) stress always tensile (c) compressive stress over part of the cycle.

The crack growth rate may now be expressed as

$$\frac{db}{dN} = f(\Delta K, R) \tag{8.4}$$

Relationships of the form given by 8.2 may still be applied, but with A and m now expressed as functions of R. The crack growth rate nevertheless shows a much weaker dependence on R than it does on the stress intensity factor range, ΔK.

A further difficulty now arises when the loading is partly compressive, as shown in fig. 8.3(c). Mode I stress intensity factors are always positive[2], since when the crack tip is closed the faces are clamped together and no stress intensity arises. For the case shown in fig. 8.3(c), the stress intensity factor range is the same as in (a) or (b), but the compressive stress applied over part of the cycle affects the growth rate, so that modified values of A and m are again needed. If the definition of R given in 8.3 is used, all cracks suffering compressive loading at some point during the loading cycle have an R ratio of zero and the parameter is not particularly useful in describing growth rate under these conditions. In these circumstances an alternative definition of R is employed, namely

$$R = \frac{\sigma_{\min}}{\sigma_{\max}} \tag{8.5}$$

[2]Occasionally negative K_I values are quoted in the literature. The only sense in which they can be valid is when they arise from one part of the loading only (e.g. a residual stress component). The net K_I arising from the total loading must neccessarily always be positive. This superposition approach is valid only as long as the *entire* crack remains open (see Chapter 6).

This definition is, of course, consistent with equation 8.3 for cases of uniform far field tensile loading. A difficulty arises, however, in the case of non-uniform loading such as that occurring in fretting fatigue. In such cases R (as defined by equation 8.5) will vary from place to place and it is not obvious at which point to evaluate it. There is, as yet, no consistent definition of R for cracks under non-uniform far field loading which are closed at the tip during part of the cycle. This is, of course, precisely the situation which pertains in many cases of fretting fatigue and the use of crack growth data obtained from experiments conducted under different conditions is made more difficult than in cases where the crack remains open. A number of different empirical relationships to enable crack growth data measured at one R ratio to be used to predict crack growth behaviour at a different R value are in common use. One approach adopted is to calculate an effective stress intensity factor range, ΔK_{eff}, as an empirical function of the actual ΔK and R. For example, Walker (1970) gives

$$\Delta K_{\text{eff}} = \Delta K(1 - R)^{n-1} \tag{8.6}$$

where n is a material constant ($0 < n < 1$). Thus, for a crack in Stage 2 growth it is possible to predict crack growth from a knowledge of the stress intensity factor range, although it is not clear how to correct for the compressive part of the stress cycle if the crack suffers closure.

Another useful feature of the $\frac{db}{dN}$ versus ΔK relationship (fig. 8.2) is that many materials exhibit a threshold value for ΔK, (ΔK_{th}) below which no growth takes place. In fatigue under uniform far-field loading, the stress intensity factor may be calculated from an expression of the form

$$K = Y\sigma\sqrt{\pi b} \tag{8.7}$$

where Y is a geometric factor. As a consequence, K increases with the square root of crack length and thus, if a constant cyclic stress amplitude is applied, cracks suffer a monotonically increasing ΔK and continue to propagate at an increasing rate. The same is not necessarily true of fretting fatigue; here cracks grow under the influence of a localised contact stress field. In such a non-uniform stress field, the factor Y in equation 8.7 becomes a function of crack length as well as geometry and will fall as the crack emerges from the high stress region surrounding the contact. If the bulk stress level is relatively low, this effect may well mean that the stress intensity factor *falls* with increasing crack length. Under cyclic loading, ΔK therefore reduces as the crack grows away from the contact and the propagation rate is reduced. In some circumstances ΔK can fall below ΔK_{th} so that crack growth ceases and the phenomenon known as self-arrest takes place.

The discussion so far in this chapter has outlined methods for predicting crack growth *rates* under fatigue loading. In a complex multiaxial stress field such as that existing in fretting, it is also necessary to predict the *direction* of crack growth. As mentioned above, there is much experimental evidence to support the assertion that cracks grow in Stage 2 primarily under Mode I loading. Several attempts have been made to predict the direction of growth from applied mechanics considerations. Erdogan and Sih (1963) proposed that crack extension was likely to take place in a direction corresponding to the maximum hoop stress $\sigma_{\theta\theta}$ ahead of the crack tip. The required stress is given by equations (6.1) to (6.4), and for the case of in-plane loading (Modes I and II only) can be expressed as

$$\sigma_{\theta\theta} = \frac{1}{\sqrt{2\pi r}}\left(K_I \cos^3\frac{\theta}{2} - 3K_{II}\sin\frac{\theta}{2}\cos^2\frac{\theta}{2}\right) \tag{8.8}$$

where θ is measured from the projection of the existing crack as shown in fig. 6.1. The angle of crack extension $\hat{\theta}$ is then found by maximising this expression. For pure Mode II loading the value of $\hat{\theta}$ obtained is 70.5^o. Other similar predictions can be obtained by determining the direction in which the strain energy density is a minimum (Sih, 1974) or by considering small crack branches from the original crack tip (e.g. Hourlier et al., 1982). These models support the experimental observations that, even if a crack initially suffers an appreciable Mode II loading, it turns quite rapidly to propagate in pure Mode I.

The predictions discussed above are based on linear elastic fracture mechanics. Inevitably, cracks will suffer some crack tip plasticity but, as discussed above, in many cases the process zone at the tip of the crack will be quite small and conditions will be characterised by the surrounding elastic stress field which is described by the stress intensity factors. Situations in which this assumption is valid are said to satisfy the conditions of 'small-scale yielding'. Under some circumstances, however, such as small cracks under high remote stresses, the crack tip plastic zone may be of appreciable size when compared to the crack length. If the size of this zone exceeds that of the region in which the elastic stresses are accurately described by the asymptotic results of equations (6.1) to (6.4), it is no longer true that the crack-tip plasticity is controlled by the elastic stress intensity factors. The conditions for small-scale yielding no longer apply and the use of linear elastic fracture mechanics is inappropriate. Estimates of the plastic zone size can be obtained from a number of models. Irwin (1960) considered a circular plastic zone ahead of the crack and obtained its radius r_p by applying Von Mises' yield criterion to the elastic stress field to give

$$r_p = \frac{1}{3\pi}\left(\frac{K_I}{\sigma_y}\right)^2 \tag{8.9}$$

for pure Mode I loading in plane strain, where σ_y is the yield stress of the material in uniaxial tension. Since the model is based on the elastic stress field, the prediction given is applicable under small-scale yielding conditions. A more sophisticated model developed by Dugdale (1960) gives a similar result:

$$r_p = \frac{\pi}{8}\left(\frac{K_I}{\sigma_y}\right)^2 \tag{8.10}$$

In applying linear elastic fracture mechanics to predictions of component life it is important to check that the requirements of small-scale yielding are met. Under conditions of significant plasticity, the techniques of elastic-plastic fracture mechanics are more appropriate to the prediction of crack propagation. These will not be discussed here, but an excellent introduction is given by Suresh (1991).

Another situation where the prediction of crack growth rate on the basis of linear elastic fracture mechanics is inappropriate, is for very small cracks which cross only a small number of grains of the material. At this scale, the representation of the cracked body as an isotropic elastic continuum is invalid, as the crystallographic arrangement of individual atoms will lead to anisotropy within each grain. Different grains will have different orientations of potential fracture planes and grain boundaries will therefore represent significant obstacles to crack growth, as the crack will have to adjust from one favourable propagation direction to another. Under such conditions, the size and nature of the obstacles need to be taken into account, and there is a considerable body of literature on the analysis of such 'short cracks' (e.g. Miller, 1991). Once again the appropriateness of linear elastic fracture mechanics must be carefully considered before it is applied to a particular problem. In fretting fatigue, the validity of LEFM may be assessed by comparing the relative scales of the crack length, contact dimension, and material microstructure. For large contacts and fine microstructure, elastic fracture mechanics may well describe the propagation of cracks for the majority of component life, and these circumstances can be shown to occur in many engineering situations.

8.2 Analysis of fretting fatigue cracks

Fretting fatigue cracks propagate under the influence of a contact stress field which, as has been demonstrated in Chapter 2, is limited in extent, having dimensions of the same order as the contact itself. The growth of a fretting fatigue crack in this region will clearly be dominated by this stress field. Once the crack has grown beyond the immediate neighbourhood of the contact, the influence of the contact stress field will have diminished to such a level that the crack is propagating mainly under the influence of any bulk stresses caused by far field loading, and the situation will be essentially one of plain fatigue. The analysis of crack propagation in fretting fatigue is therefore concerned with the portion of crack life from the time when the crack is longer than a few grain sizes until it is longer than a few contact dimensions. In many situations this will encompass a considerable proportion of component life, but there may be occasions (such as when the grain size is large, or the contact dimension small) where a linear elastic analysis of crack propagation under the influence of the contact stress field is inappropriate. In this section we will consider cases where LEFM can be applied; for these configurations the problem of analysing crack propagation reduces to that of determining the variation of stress intensity factors at the crack tip during cyclic loading. The rate of crack propagation can then be estimated using Paris Law data for the material (8.2). A basic assumption of linear elastic fracture mechanics is that crack propagation, which is determined purely by conditions at the crack tip, can be characterised by the stress intensity factors alone. It is therefore impossible for the crack to respond to any peculiarity of surface conditions, except via the stress intensity factors, and fretting fatigue simply becomes plain fatigue under a complex stress field.

A vast number of possible methods are available for the determination of stress intensity factors, and many of these have been applied to fretting fatigue. Some of the most popular methods were discussed in Chapter 6, viz.

(i) Finite element analysis

(ii) Boundary element analysis

(iii) Green's function methods

(iv) Dislocation density and other related strain nuclei techniques

Before determining the most appropriate method for application in a given problem it is helpful to consider a number of features of the fretting problem which render determination of stress intensity factors more demanding than for most other crack problems.

(i) Fretting fatigue cracks exist in a rapidly-varying contact stress field. The chosen method must be able to cope with the steep stress gradients which exist in the material through which the crack grows.

(ii) The normal contact load leads to compressive stresses, so that at some stage of the loading cycle crack closure is likely to occur. As explained in Chapter 6, such closure does not necessarily take place at the crack tip, and some means of detecting and accounting for crack face closure is needed.

(iii) Fretting fatigue is primarily a near-surface phenomenon, and it is in this region that residual stresses caused by machining processes or surface treatment predominate. These stresses can be quite high, and may well influence crack propagation since, although they do not vary cyclically, they will affect the degree of crack closure and the R-ratio experienced by an open crack.

(iv) The near surface nature of fretting fatigue means that, in many cases, only a single component boundary affects crack tip conditions during the portion of the crack's life in which fretting plays a significant part. This means that a degree of simplification can sometimes be achieved without loss of accuracy by employing a half-plane or half-space formulation.

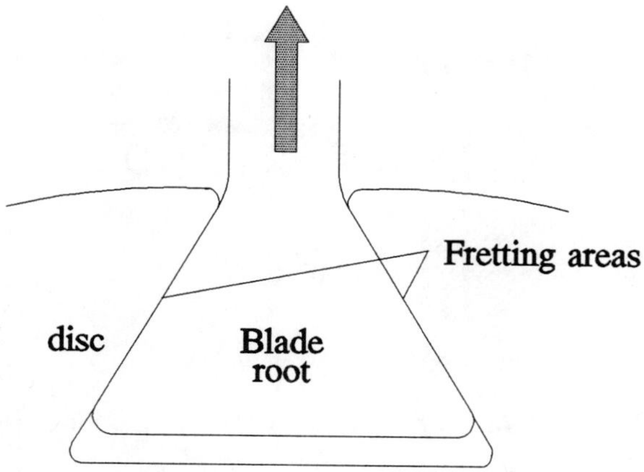

Figure 8.4: A dovetail fan blade root.

(v) Experimentally, fretting fatigue cracks are often found to exhibit a sharp kink at some point along their length, and it should be possible to incorporate this feature into the analysis.

Finite element and boundary element methods for crack analysis are, of course, special applications of more general stress analysis techniques, and it would be inappropriate to discuss them in detail here. Although they are frequently used to analyse fretting problems, there are a number of complications which can render them cumbersome or difficult to use. First, large numbers of elements are required to model the steep stress gradients. Secondly, it is normal to analyse the contact problem at the same time as the crack, and although the more powerful commercial codes now have some contact capability, there are still considerable difficulties in analysing frictional contact. Residual stresses and crack closure can be incorporated into the analysis with care (Wilks, Nowell and Hills, 1993), but the general picture which emerges is that use of these methods leads to large models which are time consuming to formulate and solve. If parametric studies are required (for example the investigation of different crack lengths, positions, or angles), the finite element method becomes particularly unwieldy, as re-meshing of the entire domain is required for each change in geometry. Thus, whilst finite element methods can be employed in most situations, albeit with very complex models, they are not necessarily the most efficient means of solution. The main strength of the method is its ability to analyse configurations situations where it is not possible to use half-plane approximations (such as analysis of a dovetail fan-blade root, fig. 8.4).

Green's function methods rely on establishing the stress intensity factors caused by a point force acting at a range of locations (e.g. along the crack faces or on the free surface) in a particular geometrical configuration (e.g. fig. 8.5). The stress intensity

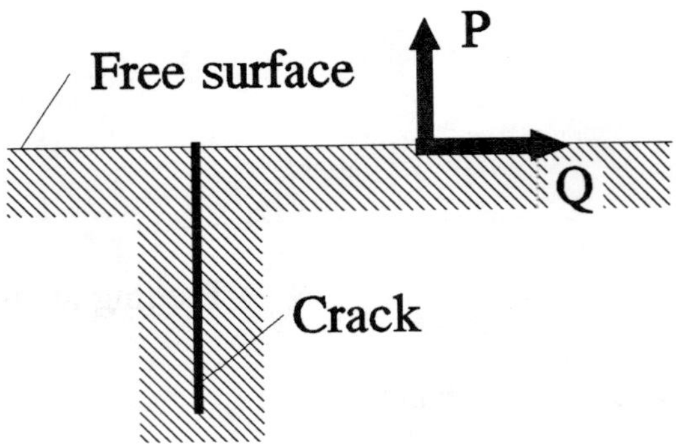

Figure 8.5: Typical geometry for which Green's functions are available.

factor K may be written as

$$K = g(x)P(x) \tag{8.11}$$

where $g(x)$ is the Green's function which needs to be established by some standard method of crack analysis. Once the Green's functions for a particular geometry have been established, the stress intensity factors due to distributed loading may be readily evaluated by employing superposition so that

$$K = \int g(\xi)p(\xi)\, d\xi \tag{8.12}$$

This approach is equally appropriate for mode I and mode II loading. The integration required in (8.12) is often quite straightforward, and can be carried out using standard numerical quadratures so that the method is superficially attractive. However, there are a number of disadvantages. Unless Green's functions are already available for the particular geometry to be analysed, it is necessary to establish them first, and this process is more laborious than solution of the problem itself. Secondly, the method relies on the principle of superposition, and addresses the case when the crack faces are traction-free. This assumes that no crack face closure occurs (see §6.3), whereas in fretting fatigue some closure is very likely at some point in the loading cycle. Thus the Green's function approach is less useful than it might first appear. Although there is no reason why the method should not be used for three-dimensional cracks, the resulting complexity renders it too cumbersome to be of practical use. Green's functions for fretting fatigue problems have been presented by Rooke and Jones (1979) and Rooke, Rayaprolu, and Aliabadi (1992). Some discussion on this latter paper is given by Hills and Nowell (1993).

The fourth approach to be considered is that introduced in Chapter 6, where distributed arrays of dislocations are employed to modify the stress field present in the uncracked body, and to impose the correct crack-face boundary conditions. The approach has been fully described in §6.2 and constitutes an extremely powerful tool

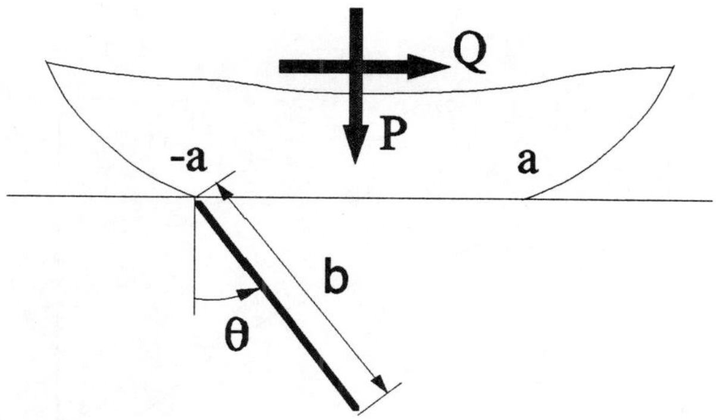

Figure 8.6: Cylinder on flat fretting fatigue configuration.

in the analysis of fretting fatigue, since it can readily cope with steep stress gradients, crack face closure, and residual stresses. Although primarily a half-plane technique it may be modified to incorporate additional free boundaries by formulating a hybrid dislocation density/boundary element method (Sturt, Nowell and Hills, 1993). The dislocation density approach is limited to plane two-dimensional configurations, but an analogous three dimensional technique is available in the form of the eigenstrain method (see §6.4). Both dislocation density and eigenstrain methods have been successfully applied to the analysis of fretting fatigue.

8.2.1 Two-dimensional analysis

Although many fretting fatigue configurations are fully three-dimensional, it is common to design experimental apparatus to give a largely two-dimensional state of stress as this clearly simplifies subsequent analysis. Nowell, Hills, and O'Connor (1987) presented an analysis of the two-dimensional configuration shown in fig. 8.6 using the dislocation density method. The experiments concerned were run in the partial-slip regime so that the prevailing traction distributions were as described here in §3.2, but modified for the presence of a bulk stress (§3.5, equations 3.58, 3.59, 3.63), so that, at the point in the cycle when maximum fretting force is applied,

$$p(x) = -p_0\sqrt{1-(x/a)^2} \qquad |x| < a \tag{8.13}$$

$$q(x) = fp_0\sqrt{1-(x/a)^2} \qquad -a < x < e-c, \quad e+c < x < a \tag{8.14}$$

$$q(x) = fp_0\sqrt{1-(x/a)^2} - fp_0\frac{c}{a}\sqrt{1-((x-e)/c)^2} \qquad |x-e| \leq c \tag{8.15}$$

where $c/a = \sqrt{1-|Q/fP|}$ (3.19) and $e = \sigma_0 a/(4fp_0)$ (equation 3.62). These traction distributions are all elliptical in form, so that the subsurface state of stress in an uncracked body may readily be found by employing superposition of suitably-scaled

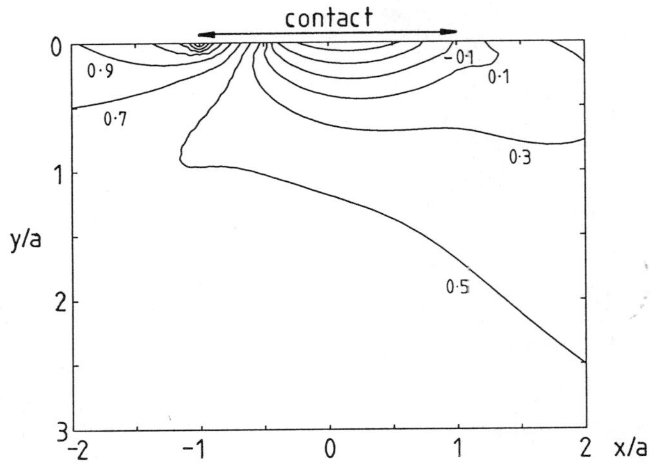

Figure 8.7: Contours of maximum principal stress, σ/p_0, for the fretting configuration shown in fig. 8.6, $Q/fP = 0.6$, $\sigma_0/p_0 = 0.6$, $f = 0.75$.

expressions for the stresses in two-dimensional Hertzian contact as given in Chapter 2. Equations (2.40) - (2.43) give the required complex potentials which may be substituted into (2.26) and (2.27) to obtain the stress components. Alternatively, the real expressions (2.44) may be used to reveal the stresses directly. The bulk stress in the fretted body, σ_0, may now be added, giving the total stress state in the uncracked body. This stress field is, in itself, a useful aid to the understanding of fretting fatigue. Figure 8.7 shows contours of the maximum principal stress[3] in the region surrounding the contact for a typical fretting configuration. The high concentration of tensile stress at the trailing edge of the contact is immediately apparent. This is the region where cracks are found to initiate experimentally. Figure 8.8 is a corresponding trajectory plot giving the (orthogonal) directions of the principal stresses. Although the directions given are those occurring in an uncracked body, they give some indication of the likely directions of crack propagation. Thus, cracks emanating from the trailing edge of the contact will be likely to propagate almost normal to the free surface, perpendicular to the direction of the maximum positive principal stress.

Knowledge of the stress field in the uncracked body enables the stress intensity factors to be determined by the method of distributed dislocations outlined in Chapter 6. The stresses along the line of the crack are evaluated and transformed into the same co-ordinate system as the crack, so that the unsatisfied normal and shear tractions, $\sigma_T(\hat{x}), \tau_T(\hat{x})$ on the crack face are established. If the crack is fully open, the net values

[3]By 'maximum principal stress' we mean the in-plane principal stress which gives the largest tension. If both principal stresses are compressive, then the least compressive is taken.

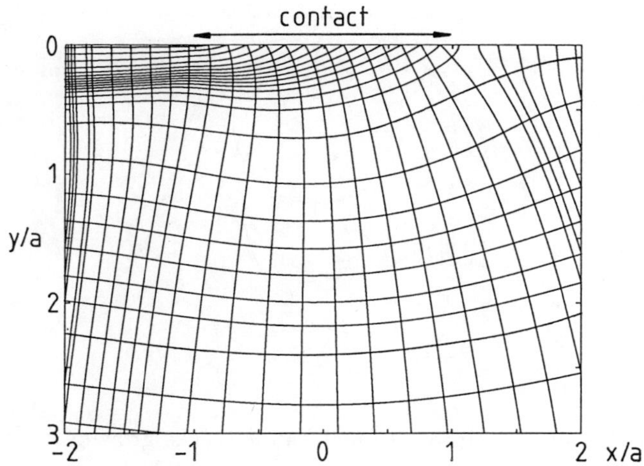

Figure 8.8: Principal stress directions for the fretting configuration shown in fig. 8.6, $Q/fP = 0.6$, $\sigma_0/p_0 = 0.6$, $f = 0.75$.

of these traction components must be zero, giving two coupled integral equations in the opening and shear components of the dislocation density (6.40, 6.41). These may be solved using the techniques presented in §6.2 to give the dislocation densities required to model the crack and, from these, the stress intensity factors (6.46). Figure 8.9 presents the results of such a calculation in the form of a polar plot. Mode I and mode II stress intensity factors were evaluated for cracks growing at a range of different angles from the trailing edge of the contact under the stress field depicted in fig. 8.7. The results for three different crack lengths are portrayed and they exhibit broadly similar features. As expected from the stress field in the uncracked body, the mode I stress intensity reaches a maximum for cracks growing almost normal to the free surface. Mode II stress intensity factors show two maxima: one for cracks growing oblique to the surface and away from the contact and a second, larger in magnitude, for cracks propagating obliquely under the contact itself. However there is a range of angles under the contact for which crack closure occurs and K_I falls to zero. An analysis of cracks growing in this region would require incorporation of crack closure as described in §6.4. In reality it is unlikely that cracks will propagate in this region since compressive stresses will clamp the crack faces together, and frictional tractions will inhibit mode II loading. Real crack faces will rarely be smooth and shear mode loading will be inhibited by interlocking of surface irregularities in the absence of crack opening.

Experimentally cracks are often observed to grow for some distance obliquely under the contact (e.g. Sato, Fujii, and Kodama, 1986). It is clear from the foregoing

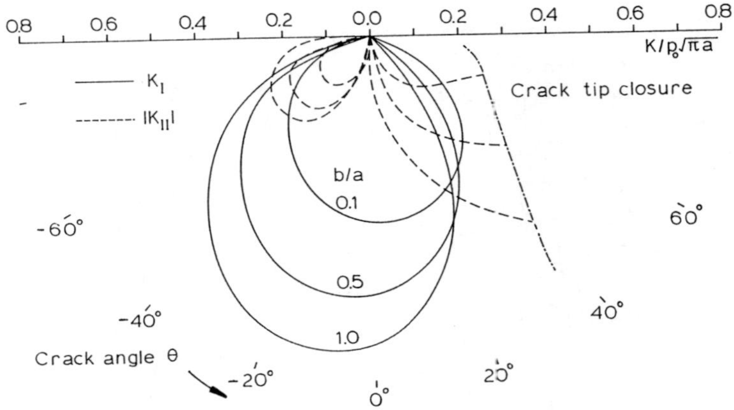

Figure 8.9: Polar plot of stess intensity factors for cracks growing at various angles from the trailing edge of a cylindrical contact (fig. 8.6), $Q/fP = 0.6$, $\sigma_0/p_0 = 0.65$, $f = 0.75$. Note that K is normalised with respect to the contact semi-width, a, rather than the more usual crack length, b.

stress analysis that stage 1 propagation under shear must be responsible for this phase of crack growth. Once cracks have turned to propagate in stage 2 (mode I), the experimentally observed growth direction normal to the surface agrees with that predicted by the analysis. Fretting fatigue cracks can sometimes exhibit quite a sharp kink on transition to mode I propagation. Kinked cracks can be analysed by the dislocation density method quite easily, although it is necessary to incorporate the stress singularity present at the kink into the solution (Hills and Nowell, 1990).

Another useful way of presenting the results of the analysis is to show the variation of stress intensity factor with crack length. A typical plot of K_I against crack length b is shown in fig. 8.10. Also shown is the stress intensity factor arising from the bulk stress alone (which is, of course, simply proportional to $\sqrt{b}$). The values depicted represent the maximum K experienced during a loading cycle. In may cases the crack will experience tip closure at some other stage in the cycle so that $K_{\min} = 0$ and thus a propagating crack will be subjected to a stress intensity factor range, ΔK, equal to $K_{\max}$. If growth is governed by a Paris Law relationship (8.2) then it would be anticipated that crack growth rate will be enhanced during the early part of the crack's life before falling to values close to those experienced under plain fatigue. Experimental measurements of crack growth rate are in good agreement with these predictions (e.g. Sato, Fujii, and Kodama, 1986).

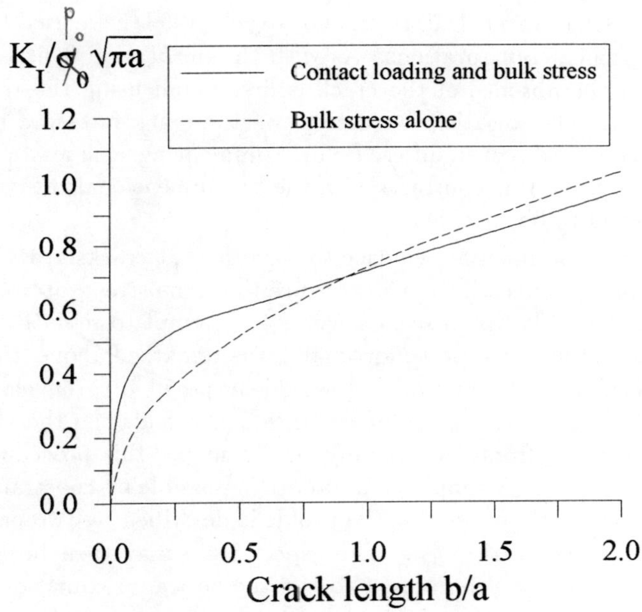

Figure 8.10: Variation of stress intensity factor, K_I, with crack length, b, for a normal crack growing from the trailing edge of a cylindrical contact. $Q/fP = 0.6$, $\sigma_0/p_0 = 0.65$, $f = 0.75$. K is normalised with respect to contact semi-width rather than crack length.

8.2.2 Three-dimensional analysis

Whilst two-dimensional analysis of cracks is attractively straightforward, many fretting geometries are fully three-dimensional. Even in a nominally two-dimensional stress field such as that under cylindrical contact, cracks will initiate as thumbnails and propagate for some distance before coalescing to form a through thickness crack. We must therefore analyse three-dimensional cracks, either to study small flaws in a two-dimensional stress field or to understand crack behaviour in a three-dimensional configuration such as a sphere on a flat (which is attractive from an experimental point of view, see Chapter 7). Dai, Hills, and Nowell (1993) have used the eigenstrain method to study such configurations. As with the dislocation density method, the state of stress in the absence of the crack is first found using the results given in Chapters 2 and 4. The unsatisfied tractions on the crack faces are then cancelled by use of displacement discontinuities. Before commencing such an analysis we must choose the crack *shape* (this contrasts with the two-dimensional case where only the crack length needed to be specified).

There is much experimental evidence to suggest that cracks grow in such a way that all points along the crack front experience the same stress intensity factor (e.g. Soboyejo et al., 1989). It would seem sensible to attempt to replicate this so called iso-K behaviour in the analysis. Unfortunately, as remarked above, the crack shape needs to be specified first and whilst it would be possible to develop an iterative method by making an initial guess of crack shape, calculating the stress intensity factors along the crack front, and modifying the shape, this procedure is likely to prove tedious in practice. A simplified approach is possible by constraining the crack to be semi-elliptical in shape so that the profile is described by two parameters (the depth b and the aspect ratio c/b). The aspect ratio may then be adjusted for a particular value of crack depth in order to find the best approximation to a constant K profile. One possible way of doing this is to compare values of K at the bottom of the crack with those where it breaks the free surface. However, it is necessary to take care at the surface since the singularity in stress deviates from the normal $\frac{1}{\sqrt{r}}$ value (Benthem, 1977) and it will be difficult to establish convergence unless this effect is taken into account.

Figure 8.11 shows the results of an analysis carried out by the eigenstrain method for a typical case of a semi-elliptical crack growing from the trailing edge of a cylindrical Hertzian contact, under the influence of the contact stress field, together with a uniform far-field tensile stress, σ_0 . The figure shows the variation of Mode I stress intensity factor, K_I, around the crack front for several different values of ellipticity, c/b. It is clear that a reasonable approximation to an iso-K crack front can be obtained with a simple semi-elliptical shape (in this case $c/b = 1.5$) under this contact stress field, although a better approximation would, of course, be possible by using more parameters to model the crack shape. Calculations of this kind enable the most likely crack shape at any particular stage of its growth to be determined. It is then

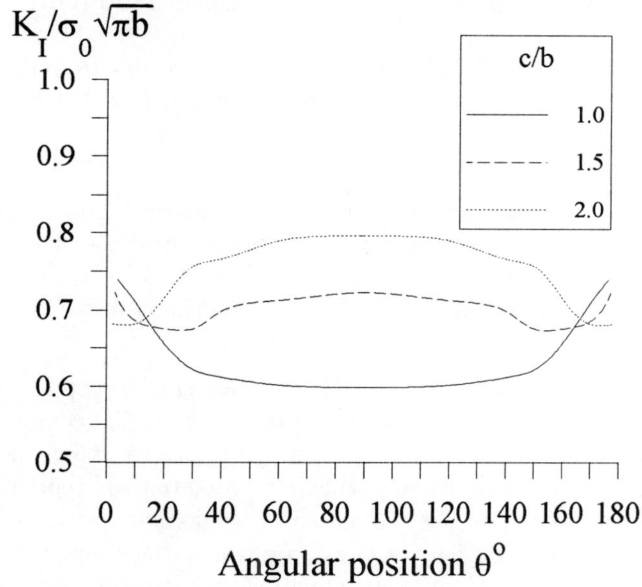

Figure 8.11: Variation of Mode I stress intensity factor with angular position along the crack front for a semi-elliptical crack normal to the free surface and emanating from the trailing edge of a cylindrical Hertzian contact of semi-width a; $Q/fP = 0.75$, $\sigma_0/p_0 = 0.33$, $f = 0.75$, crack depth $b/a = 1.0$.

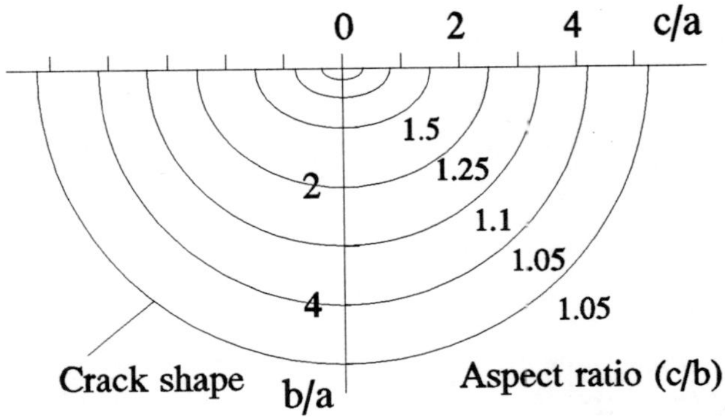

Figure 8.12: Evolution of iso-K crack shape for a crack growing from the trailing edge of a cylindrical Hertzian contact; $Q/fP = 0.75$, $\sigma_0/p_0 = 1.0$, $f = 0.75$.

possible to examine the evolution of crack shape as the crack grows. Typical results are shown in fig. 8.12, again for a crack growing from the trailing edge of a cylindrical Hertzian contact. It can be seen that there are considerable variations in ellipticity during crack growth; these are caused by the non-uniform nature of the contact stress field.

Similar analyses can, of course, be carried out for cracks propagating under the influence of other contact geometries, including the sphere on flat configuration. Some typical results are displayed in fig. 8.13 for this geometry. As might be anticipated, cracks with a high aspect ratio (i.e. very wide in relation to their depth) have a higher stress intensity factor at the deepest point than they do close to the surface, so that they might be expected to propagate away from the surface into the material and so reduce their aspect ratio. Conversely, cracks with low aspect ratios experience a higher stress intensity factor at points on the crack front close to the free surface and might therefore be expected to grow wider rather than deeper. An estimate of the iso-K crack shape may be obtained by determining the aspect ratio where the surface and bottom values of the stress intensity factor are equal.

These results demonstrate that the eigenstrain method is a powerful tool for crack analysis under fretting conditions. By comparison with the finite element approach, it is capable of producing much more rapid solutions for stress fields resulting from the well-defined contact geometries which are to be preferred for experimental work. A particularly attractive feature is that a range of solutions for different sizes of cracks (of the same geometrically similar shape) can be evaluated without the need to re-mesh the crack face or re-calculate the kernel of equation (6.71). For each different size of crack the contact stress field is re-scaled and re-evaluated at the chosen collocation points. This simply changes the left-hand side of equation (6.71) and solution of the discretised equation is straightforward and rapid.

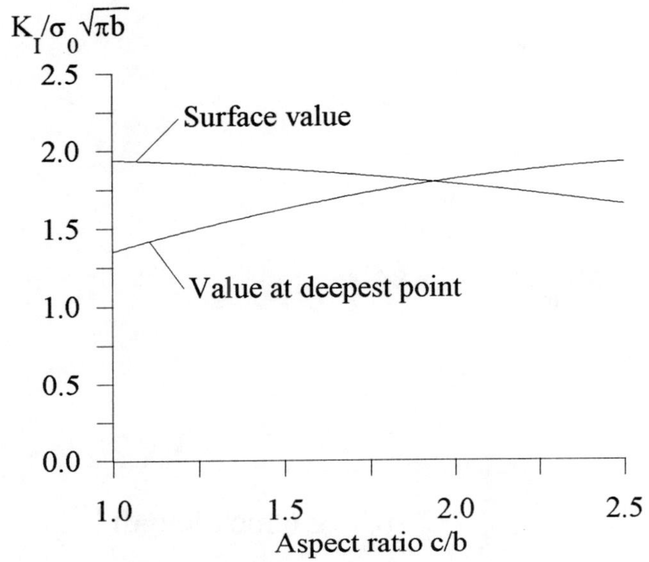

Figure 8.13: Variation of Mode I stress intensity factor with crack aspect ratio for a semi-elliptical crack normal to the free surface emmanating from the trailing edge of a spherical Hertzian contact. Salient parameters compare with those used by Kuno et al. (1989); $Q/fP = 0.89$, $\sigma_0/p_0 = 0.33$, $f = 0.7$, $b/a = 0.5$.

8.3 Crack arrest in fretting fatigue

One interesting and practically significant feature of fretting fatigue is that in many situations cracks may be formed and propagate for some distance before stopping. If this occurs the cracks will not cause overall component failure and the design will be safe. It is therefore important to study conditions for crack arrest in addition to those for initiation and propagation. A crack in a uniform stress field experiences a stress intensity factor of the form

$$K = Y\sigma\sqrt{\pi b} \tag{8.16}$$

where Y depends only on crack geometry. For cracks propagating from the surface of a half-plane (i.e. in cases where all other free boundaries are remote), the only geometric length scale is the length of the crack itself, and therefore Y is a constant. If other boundaries are present (e.g. in the case of a crack propagating through a strip of material) Y frequently increases with crack size as the tip approaches the remote boundary. Thus most cracks occurring in engineering components experience a stress intensity factor which increases with crack length faster than $\sqrt{b}$. Hence, under cyclic loading, the stress intensity factor range increases monotonically with crack size (fig. 8.14). A crack of a certain size, b_0, is needed in order to give rise to a stress intensity factor range ΔK larger than the threshold value ΔK_{th}. Such a crack can either be

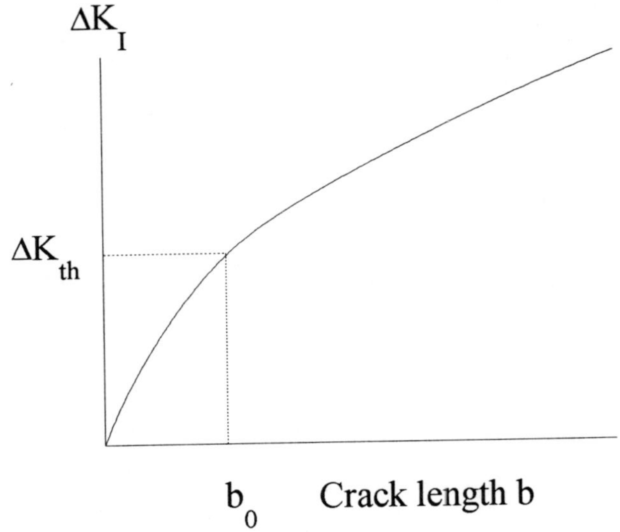

Figure 8.14: Variation of stress intensity factor range ΔK for a crack in a cyclically-varying tensile stress field of amplitude σ_0, plotted on an absolute (i.e. not normalised) scale.

a pre-existing flaw or can arise as the result of an initiation process but, once it is present, ΔK will always remain above the threshold value and propagation will take place in the linear-elastic fracture mechanics regime until failure commences.

Fretting fatigue cracks can exhibit an entirely different behaviour. They start at a point of extremely high local stress, giving rise to high stress intensity factors for short cracks. The contact stress field is, however, extremely localised, and its influence will decay quite rapidly as the crack grows away from the contact. Further crack propagation effectively takes place under the influence of the far-field stresses alone, and if these are small compared with the contact stresses, the crack may experience a fall in ΔK as it grows away from the contact. In certain circumstances the stress intensity factor range may fall below the threshold value, and if this occurs, crack growth will cease. Figure 8.15 displays the variation of stress intensity factor with depth for cracks growing normal to the free surface from the trailing edge of a cylindrical Hertzian contact. Note that K is plotted normalised with respect to the (constant) contact semi-width, a, rather than the more usual crack length, b, in order to give a true impression of the absolute value of the stress intensity factor. Several different curves are given, corresponding to different amplitudes of cyclic far-field tensile stress, σ_0. It can be seen that lower values of σ_0 lead to a decreasing stress intensity factor range ΔK, and high values of σ_0 to an increasing range. The threshold stress intensity factor range ΔK_{th} for a particular material may be plotted as a horizontal line (e.g. fig. 8.16); if the ΔK curve crosses this line more than once

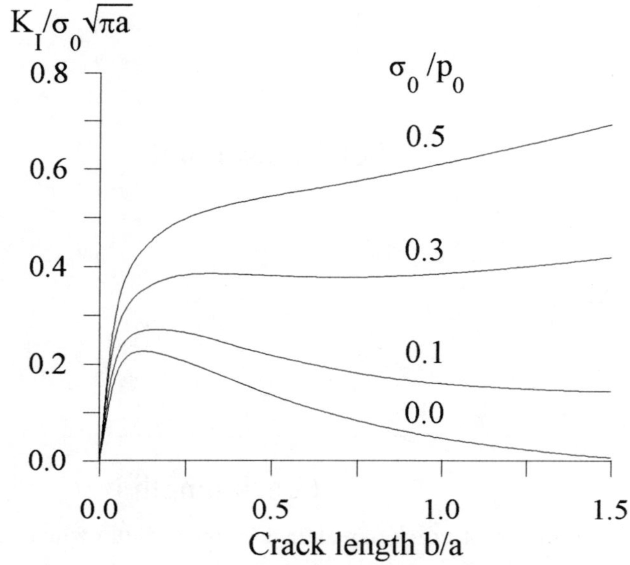

Figure 8.15: Variation of mode I stress intensity factor range, ΔK_I, with crack length, b, for a normal crack growing from the trailing edge of a Hertzian contact. $Q_{max}/fP = 0.75$, $f = 0.75$. K is normalised with respect to the contact semi-width, a.

we may be infer that crack growth will cease.

An interesting corollary of this mechanism of self-arrest is that it provides a possible explanation for the size effect reported by a number of authors (e.g. Bramhall, 1973, Hills, Nowell, and O'Connor (1988). Analysis of the Hertzian contact of cylinders (§2.3.1) shows that the peak contact pressure, p_0, is given by equation (2.38) as

$$p_0 = \frac{2P}{\pi a} \tag{8.17}$$

whereas the contact semi-width, a, is given by

$$a = \sqrt{\frac{2PA}{\pi k}} \tag{8.18}$$

where P is the normal load, A the composite compliance, and k the relative curvature of the contacting bodies. Substitution of 8.18 into 8.17 gives

$$p_0 = \sqrt{\frac{2Pk}{\pi A}} \tag{8.19}$$

It is now possible to compare equations (8.19) and (8.18) to deduce that, for a particular pair of materials a (which governs the *extent* of the contact and therefore of the

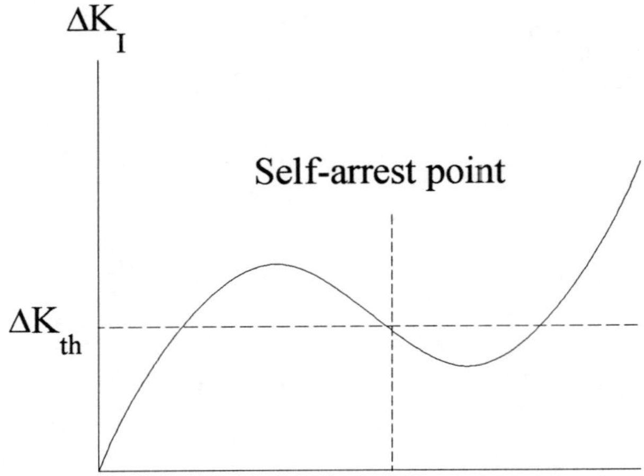

Figure 8.16: The occurence of crack arrest in fretting fatigue when ΔK falls below ΔK_{th}.

subsurface stress field) varies as $\sqrt{P/k}$, whereas p_0 (which governs the *magnitude* of the stresses) varies as $\sqrt{Pk}$. Thus, by judicious choice of P and k it is possible to produce geometrically similar contact stress fields which are identical in magnitude[4], but where the contacts differ in size so that the stress fields differ in extent. Figure 8.17 shows the variation of stress intensity factor with crack length for two geometrically similar contacts of this type. When plotted normalised with respect to the contact half-width, the two curves would (as a result of the geometric similarity) lie on top of one another, but when plotted on absolute scales, a different picture emerges. It will readily be seen that the smaller contact is liable to produce self-arrest, whereas the larger one will not.

Crack self-arrest is clearly important in practice. Severe fretting at the surface of a component may well initiate a number of micro-cracks, but if the far-field stresses are mild (or at any rate contain only a small cyclic element) they may not lead to overall failure. Such arrested cracks are found in a number of engineering components including dovetail blade roots and power station generator rotors. One interesting method of investigating the phenomenon is to employ an oscillating fretting force together with a *static* far-field load, as was done by Kuno et al. (1989). Under these circumstances the oscillatory component of the stress field decreases rapidly away from the contact. A typical plot of maximum and minimum K_I values suffered by a crack under such loading is given in fig. 8.18. The decrease in ΔK is clearly visible. Experiments of this kind will almost always produce arrested cracks, provided the

[4]i.e. the peak values of the stresses are the same.

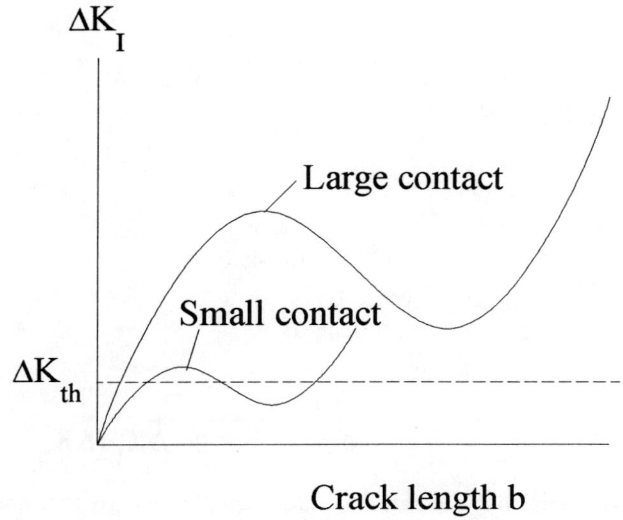

Figure 8.17: Variation of stress intensity factor with crack length for two geometrically similar contacts producing stress fields of the same magnitude but differing in extent. Co-ordinate axes are not normalised.

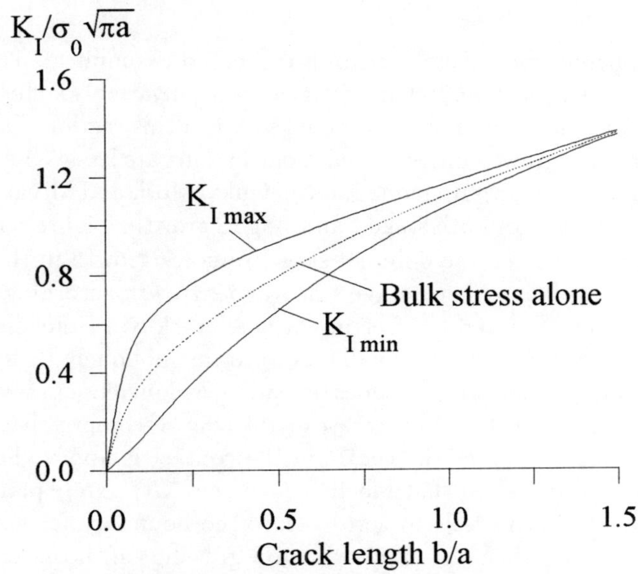

Figure 8.18: Maximum and minimum K_I values suffered by a crack growing normal to the free surface from the trailing edge of a contact under oscillatory fretting load but static bulk tension. Cylindrical Hertzian contact, $Q/fP = 0.75$, $f = 0.7$, $\sigma_0/p_0 = 1.0$. K is normalised with respect to the contact semi-width, a.

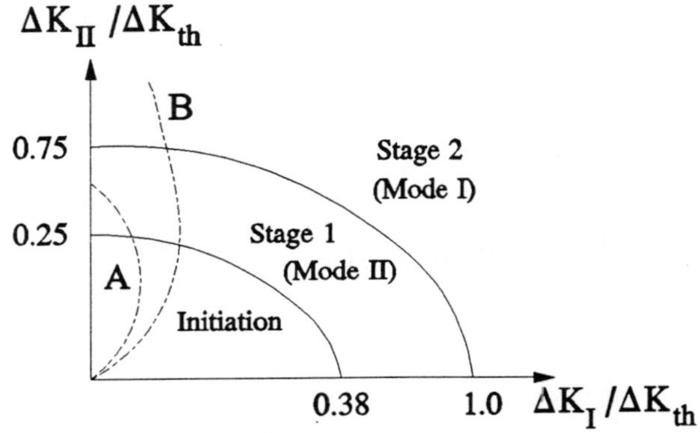

Figure 8.19: Boundaries between stage 1 and stage 2 crack growth under mixed-mode loading according to Otsuka et al. (1975).

surface fretting is severe enough to initiate them, although the arrested cracks may well be too short to be easily detected.

In reality the phenomenon of crack arrest is rather more complicated than has been presented so far. This is because fretting fatigue cracks invariably suffer mixed-mode loading rather than pure mode I, as considered so far in this section. Descriptions of the thresholds for crack growth under mixed-mode loading are less well-established for mixed-mode loading than for pure mode I, but Otsuka, Mori, and Miyata (1975) have presented possible criteria for both stage 1 and stage 2 growth. Figure 8.19 depicts the proposed boundaries between the different stages of crack growth in the $\Delta K_I, \Delta K_{II}$ plane. Thus, a crack will experience stage 1 or stage 2 growth according to the values of ΔK_I, and ΔK_{II} experienced. A fretting fatigue crack will suffer both modes of loading, as shown in 8.9. Consider a small crack growing obliquely under the contact from the trailing edge. Initially both modes of loading will increase in severity, but the mode I component will still start to decrease as the zone of compressive stresses (and consequent crack closure) under the centre of the contact is approached. This will result in a trajectory similar to that labelled A in the $\Delta K_I, \Delta K_{II}$ plane (fig. 8.19). If the crack closes (i.e. ΔK_I falls to zero) before the boundary for stage 2 growth is crossed (trajectory A) then the crack will stop growing. If, however, the stage 2 boundary is crossed (trajectory B), then the crack is likely to turn to pure mode I (i.e. approximately normal to the free surface) and propagate to failure. Once again, larger contacts are more likely to give rise to propagating cracks and smaller contacts to ones which self-arrest. An explanation of the size effect observed by Bramhall (1973) based on this type of mixed-mode threshold behaviour has been proposed by Hills,

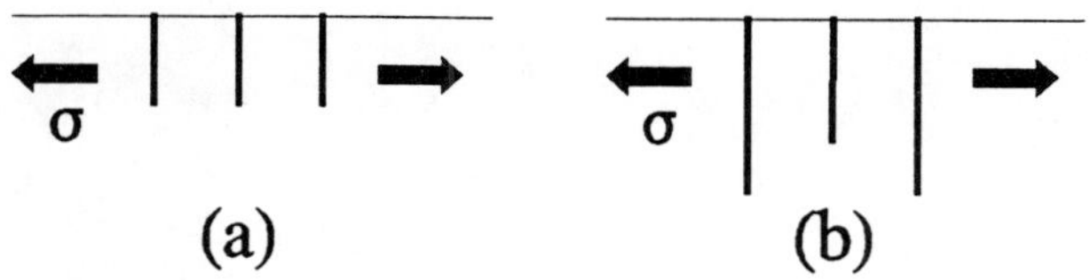

Figure 8.20: An array of three cracks under uniform far-field tension; (a) initially, (b) after several cycles of loading.

Nowell, and O'Connor (1988), although the authors now feel that differences in the surface conditions which produce crack initiation provide a more likely explanation for the range of experimental parameters considered (see Nowell and Hills, 1990).

One further mechanism of crack arrest which deserves mention is that of crack shielding. Consider the array of three cracks under uniform tensile stress shown in fig. 8.20(a). Calculations of the stress intensity factors reveal that the central crack suffers less severe loading than the two outer ones. Intuitively this might be expected, since the outer cracks disturb the loading seen by the central crack and 'shield' it from some of the load. Under cyclic loading the outer cracks experience a higher ΔK and thus propagate more quickly than the central flaw. Hence the situation shown in fig. 8.20(b) develops. Here the outer cracks are longer than the central one and as a consequence experience even higher ΔK, so that their growth accelerates still further. Eventually the ΔK for the central flaw falls below the threshold value and growth stops, leaving the two outer cracks to propagate to failure. Whilst this phenomenon is of less practical interest than the arrest mechanisms discussed earlier, since not all cracks are arrested and some still propagate to failure, it is clearly likely to occur in fretting fatigue, where the severe surface conditions give rise to the initiation of large numbers of small cracks.

Theoretical studies of this configuration have been carried out by Dubourg and Colin (1992) using the dislocation density method. Their model incorporates both crack face friction and closure. They show that alternate cracks tend to propagate so that, in an array of three cracks, numbers one and three tend to dominate; with five cracks, numbers one, three, and five dominate initially, but eventually only the two outer cracks continue to propagate.

Chapter 9

Analysis of crack initiation

9.1 Introduction

Fretting fatigue is particularly damaging because it can affect both the initiation and propagation phases of crack life. In the preceding chapter its effect on propagation was discussed and it was shown that in this phase of the crack's life the problem can be reduced to that of analysing cracks growing in a particularly severe and complex stress field. The effect of fretting on crack *initiation* is likely to be far less amenable to analysis since initiation will take place at a microscopic level. Nevertheless the area is important for the study of fretting fatigue since initiation can take up a significant proportion of crack life under some conditions. Before discussing crack initiation in fretting fatigue, however, it is appropriate to discuss the nature of crack initiation in general and to highlight a few important features.

First, it is necessary to examine what is meant by initiation as distinct from propagation. The definition will vary somewhat according to the standpoint of the observer, e.g. whether he or she is an experimentalist or a modeller, and if the former, what level of crack detection equipment is available. However a number of important observations can still be made. It is essential to note that initiation must be viewed as a continuous *process* rather than a discrete event. Crack initiation takes place over a period of time and involves the gradual accumulation of damage. Whilst it is possible to say that there is no crack (at least above a certain size) before cyclic loading commences and subsequently that a crack becomes observable after a particular number of cycles, there is no *specific* point in time at which the damage becomes a crack. A number of workers feel the term initiation is inappropriate since it does suggest a single event and the term 'nucleation' is often employed as an alternative. The term 'initiation' is, however, in such widespread use that it seems unhelpful to reject it on the grounds of semantics and it will be retained here provided that it is understood that it describes a process rather than an event.

The second important observation regarding crack initiation is that it is occurs on a microscopic scale and takes place within one (or at most two or three) grains of a polycrystalline material. Hence, any description of the initiation process will need to be rather general, and somewhat qualitative. There seems little hope of describing the *actual* initiation conditions since to do so would require knowledge of the exact local geometry, crystal orientation, and stresses at all possible initiation sites. The best that might be hoped for is to characterise the conditions in some more general way and to predict where initiation might occur and how long the process might take. There are two possible ways of developing such an understanding:

(a) A bulk or 'black-box' approach, where macroscopic variables such as stresses, strains, or slip amplitudes are calculated and correlated with the observed sites and durations of crack initiation. No attempt is made to model the actual process taking place. Instead the likelihood of initiation is described empirically as a function of a number of bulk variables. A different combination of variables is likely to be needed for each material under consideration, and no physical reasoning is applied or deduced.

(b) A local or micromechanics approach, where a model of a particular initiation mechanism is established and used to predict the likelihood of crack initiation based on a characteristic description of local conditions.

Both types of approach will be covered later in the chapter and their application to fretting problems highlighted.

A final important preliminary consideration is to define the diffuse boundary between the processes of initiation and propagation. From an experimental standpoint, initiation is often taken to encompass that period of component or specimen life from the beginning of loading until an observable flaw is detected. The remainder of the life is then defined as propagation. It is quite common to find papers in the literature which place a numerical value on 'initiation life' defined on this basis. The problem with this definition is, of course, that the smallest observable flaw size is a function of the detection method employed and of the experience of the user. Different methods of crack detection (e.g. optical, a.c. potential drop, acoustic emission) have quite different resolutions and so the quoted 'initiation life' often describes the sensitivity of the detection equipment more accurately than the behaviour of the specimen.

A mechanics approach leads to a quite different definition of the initiation period. Propagation is often taken to refer to that period of component life where the crack behaviour can be adequately described by fracture mechanics (either linear-elastic or elastic-plastic). This leaves the label 'initiation' to be applied to all that occurs before a growing flaw enters the fracture mechanics regime. As such, initiation is taken to refer to the formation of a flaw which encompasses a few grain sizes in length. Much recent work has been undertaken (see, e.g., Miller, 1991) to extend the application of fracture mechanics back into the region of short cracks. As a result, the proportion of life referred to as initiation has perhaps decreased somewhat.

The experimental and mechanical definitions of the initiation phase of crack growth may at first seem quite different. However there is a measure of agreement between the two approaches. Better experimental techniques reveal smaller and smaller flaws and enable their behaviour to be monitored. Once such data have been collected, it is possible to develop mechanical models to describe the behaviour and thus, the two definitions remain approximately in step; once an observable flaw is detected it is frequently possible to describe its behaviour by means of a fracture mechanics model (albeit one which requires modification to account for crack tip plasticity or grain boundaries, but where the scale of the geometry is at least such that ideas of continuum mechanics apply). The lack of a consistent definition of crack initiation should always be borne in mind when discussing this period of component life. In this chapter a mechanics viewpoint will be adopted so that initiation will be taken to refer to the process of accumulation of damage up to the point at which a crack exists and encompasses one or two grain sizes, so that its subsequent behaviour may be described by fracture mechanics.

9.2 Crack initiation in fretting

The remarks made in §9.1 concern crack initiation in general. There are, however, a number of features of initiation under fretting conditions which render the situation quite different from that commonly encountered in normal (or plain) fatigue. In plain fatigue, cracks can grow from a pre-existing flaw, such as a void or inclusion in a casting, in which case the location of the crack is specified by the initial flaw and there is no initiation period to the component life. Alternatively, if there is no pre-existing flaw of a suitable size, accumulation of damage must take place to form a flaw before crack propagation can take place. Such damage accumulation is clearly more likely at regions of higher stress, but in many components the stress field varies quite slowly with position so that it is not possible to predict the location of crack initiation with any certainty. In contrast, the situation in fretting fatigue is quite different. The presence of the contact imposes a severe stress concentration at the surface of the component, and fretting conditions often lead to the initiation of a number of micro cracks. The concentration of stresses is so localised that the *location* of initiated flaws can often be predicted with some considerable degree of certainty. In many cases the fretting effect is so severe that large numbers of microcracks are initiated in the fretted region. This contrasts with the case of plain fatigue where a single crack is the most likely outcome of the initiation process. Crack initiation in fretting takes place at the surface of the component. In plain fatigue this is not always the case since subsurface initiation at voids or inclusions can occur (albeit less frequently than surface-initiated flaws).

In reality, of course, the term fretting fatigue encompasses a wide range of possible conditions, ranging from mild fretting in the presence of severe bulk specimen stresses above the plain fatigue limit of the material, to heavy fretting in the presence of

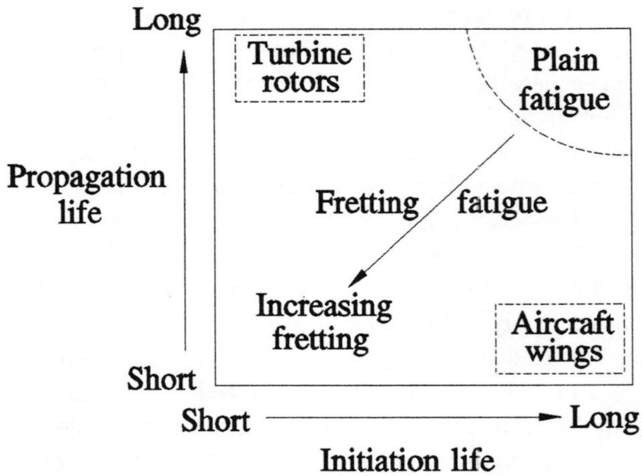

Figure 9.1: Schematic diagram showing the range of conditions which may be termed fretting fatigue.

relatively mild bulk stresses. The question 'Does fretting fatigue primarily accelerate crack initiation or crack propagation?' is frequently asked. There can be no definitive answer. The primary effect will depend on the precise conditions prevailing. It is interesting to compare a number of practical cases. Turbine rotors for power plant are frequently found to contain large numbers of non-propagating fretting fatigue cracks after relatively few cycles of loading (such as might be experienced during the initial balancing of the rotor). In this case initiation is seen to be relatively rapid whereas cracks propagate slowly or do not grow at all. In contrast, it is common in the aircraft industry to replace a component once a detectable crack is observed. In this case the initiation phase of life dominates and, for components subjected to contact loading, interest must be focused on the effect of fretting on crack initiation. A true understanding of the phenomenon of fretting fatigue demands an explanation of its influence on initiation *and* propagation so that the entire spectrum of failures encountered in practice can be addressed. Figure 9.1 shows the range of conditions which may be termed fretting fatigue. The precise location of any particular fretting problem on this diagram will depend on the actual conditions in the neighbourhood of the contact, including the local values of the bulk loading.

The analysis of crack initiation in fretting is far more challenging than that of propagation. A wide range of factors will affect the number of cycles required for damage to accumulate but they can conveniently be grouped into three categories:

(a) Material factors, including grain size, morphology, and other microstructural properties of the material in addition to 'bulk' properties such as yield stress or fracture toughness.

(b) Environmental factors, including the background temperature and whether there are any corrosive agents present.

(c) Mechanical factors such as surface tractions, amplitude of microslip, coefficient of friction etc.

A full understanding of the problem will only be gained when all these factors are brought under consideration. However the range of expertise necessary to master these three areas is considerable, and at present few effects are particularly well understood. It is beyond the scope of this book to examine all these areas in detail, and we will restrict our attention to the mechanical aspects of crack initiation. Such a restriction is not intended to imply that the other effects are unimportant. However, there is much that can be said about initiation from the mechanics standpoint alone, and it is from this perspective that we wish to start. The principal mechanical factors affecting crack initiation will be the surface (or near surface) stress state and the relative amplitude of slip between the contacting surfaces. The dependence of crack initiation on the stress state is relatively obvious, and mirrors the behaviour observed in plain fatigue, but the importance of slip amplitude is more difficult to explain; one of the principal goals of any mechanics analysis of crack initiation in fretting must be to incorporate the effects of relative slip amplitude.

The influence of slip amplitude on total fretting fatigue life has been known for may years. One of the first comprehensive investigations of the effects of slip was carried out by Nishioka and Hirakawa (1969). Their investigations concentrated mainly on the partial slip regime but other work has extended into the regime of fully-sliding contact (e.g. Spink, 1990). Vingsbo and Söderberg (1988) have summarised the effect of slip amplitude on wear rate and fretting fatigue life by use of a diagram similar to that shown in fig. 9.2. At low slip amplitudes fretting is mild and long fatigue lives are recorded. The wear rate is correspondingly low. As the amplitude of slip is increased, the fatigue life falls and the wear rate increases as a result of the increased fretting. A minimum in fatigue life is eventually reached, however, often around the transition from partial slip to full sliding. The wear rate continues to increase if slip amplitude is increased beyond this value, but the fatigue life increases again. This phenomenon of increased life is often interpreted as being a result of the rapidly accelerating wear rate. Although fretting is quite severe and cracks are readily initiated, the enhanced wear rate means that embryo cracks are worn away before they have chance to grow, and hence wear becomes the dominant manifestation of fretting.

The variation of fretting fatigue life with slip amplitude is something of a puzzle from a mechanics point of view. Once a crack has entered the propagation phase its growth will be governed purely by conditions in the process zone at the crack tip, as characterised by the stress intensity factor. Thus propagation must be governed purely by stresses. As mentioned in Chapter 8, Bramhall (1973) and Nowell (1988) investigated a size effect in the fretting of a Hertzian contact; they kept the magnitude of the stresses constant but varied the extent of the stress field. The fatigue lives

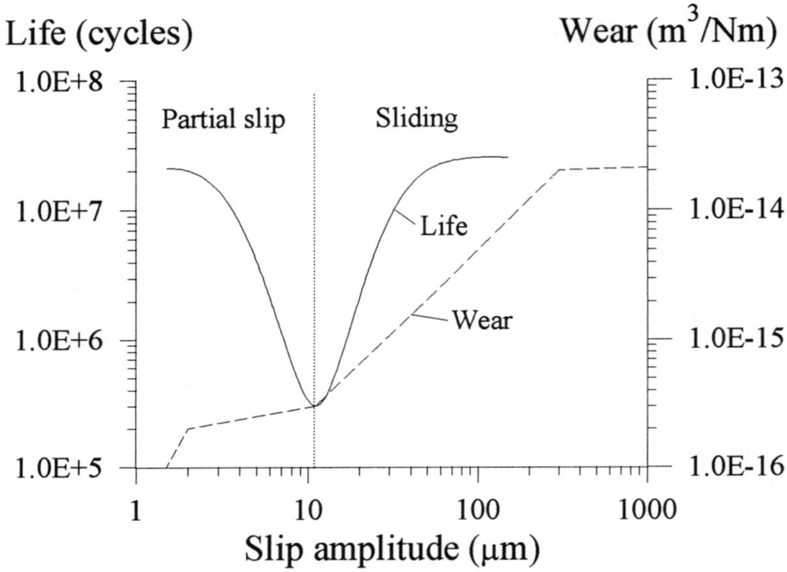

Figure 9.2: Schematic diagram showing a typical relationship between slip amplitude, wear rate, and fretting fatigue life (after Vingsbo and Söderberg, 1988).

of the specimens tested were found to be much lower with the large contacts than for the small ones. One consequence of varying the contact size in this way is that the amplitude of slip also varies (since slip in partial contact is proportional to contact size for a given tangential force coefficient (Q/fP): see, for example, equations (3.26)-(3.32)). Thus the size effect can be interpreted as another manifestation of the dependence of fretting fatigue life on slip amplitude. These experiments are, however, particularly noteworthy since the magnitude of the stresses was held constant whilst the contact size (and hence the slip amplitude) was varied. Hence it is clear that a model based on stresses alone cannot satisfactorily predict observed behaviour; slip amplitude must be included in some way.

These observations serve to highlight some of the features of a credible initiation model. Fretting fatigue life depends on slip amplitude as well as stresses, yet once the propagation phase of life has been reached, growth rate depends on stresses alone. Thus the effect of slip amplitude is on crack initiation, and must be incorporated into any initiation model developed. As mentioned in §9.1, it is possible to conceive of models based either on bulk properties or upon a characterisation of local conditions. We shall now proceed to examine models for initiation in fretting fatigue developed on the basis of both approaches.

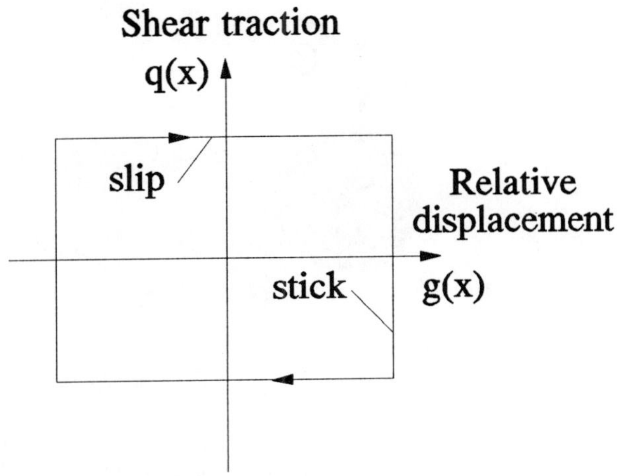

Figure 9.3: Variation of shear traction and relative displacement between surfaces at a particular point in the contact during cyclic loading.

9.3 The bulk approach to initiation analysis

Early attempts to explain the initiation of fretting fatigue cracks were based on stress criteria alone. For example, the contours of most positive principal stress depicted for a cylindrical contact in fig. 8.7 show a high stress concentration at the trailing edge of the contact and it is superficially attractive to postulate that initiation might take place here at a given value of some particular stress-derived variable such as von Mises' yield parameter or the maximum principal stress. However, it has been shown above that such an approach is bound to fail since the effects of slip amplitude are not taken into account. Progress towards a more sophisticated model was made by Ruiz and Chen (1986) as part of an experimental investigation into fretting in the dovetail roots of turbine blades. They proposed that the fretting fatigue damage occurring could be characterised by a parameter $\tau.\delta$. This variable was evaluated at all points on the contact surface as the product of the maximum shear traction obtaining during a cycle, τ, and the (local) value of the microslip between the contacting surfaces, δ. The variation of this parameter along the flanks of the dovetail root was calculated using the finite element method and displacements were also measured experimentally by Moiré interferometry. The locations of maximum damage were found to correlate quite well with local maxima of $\tau.\delta$.

Although empirical in basis, the damage parameter $\tau.\delta$ has a physical interpretation. If the variation of shear traction and relative slip is plotted *for a particular point in the contact* during cyclic loading, the rectangular hysteresis loop shown in fig. 9.3 emerges within the slip zones. No slip takes place until the shear tractions reach the limiting value $+fp(x)$. Slip then starts and the value of shear traction re-

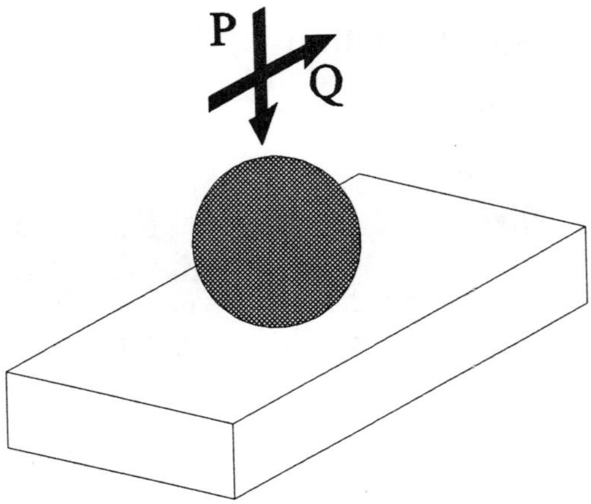

Figure 9.4: Sphere on flat fretting geometry.

mains constant until the loading phase of the cycle stops and unloading commences. Instantaneous adhesion occurs at this point and no further slip takes place there until the shear tractions reach $-fp(x)$. The energy expended in one particular cycle is given by the area of this loop which is, of course, proportional to $\tau.\delta$. The damage parameter therefore gives a measure of the density of frictional energy dissipation at a particular point in the contact. It is not unreasonable to suggest that, whatever the mechanism of damage accumulation, the amount of damage will increase in line with the amount of energy supplied and hence it would appear reasonable to use $\tau.\delta$ as an empirical measure of fretting damage.

Although the damage parameter is useful, there is not necessarily a good correlation between the sites of maximum fretting damage and the location of crack initiation. Consider the sphere on flat configuration depicted in fig. 9.4. The contact patch is circular and, under oscillatory tangential loading, small cracks develop at the leading and trailing edges of the contact (see Kuno et al., 1989). Values of the damage parameter can be calculated using Mindlin's analysis (Mindlin 1949) presented in §3.3. Both the shear tractions and the relative displacement component in the direction of loading are functions of radius alone, so that the product $\tau.\delta$ gives an axisymmetric distribution of the damage parameter (fig. 9.5). Although Mindlin's analysis is not exact for materials exhibiting a Poisson effect[1], more accurate analyses show only a modest departure from the axisymmetric distribution of $\tau.\delta$ (Munisamy, Hills, and Nowell, 1992). In practice, damage is found to be more or less uniform around the slip annulus (see Johnson, 1955), but as remarked above, cracks start in the slip zones at the leading and trailing edges of the contact. The explanation for this is that more than just surface damage is required in order to initiate a crack. In

[1]In reality there is a small transverse component of relative displacement, see §3.3.

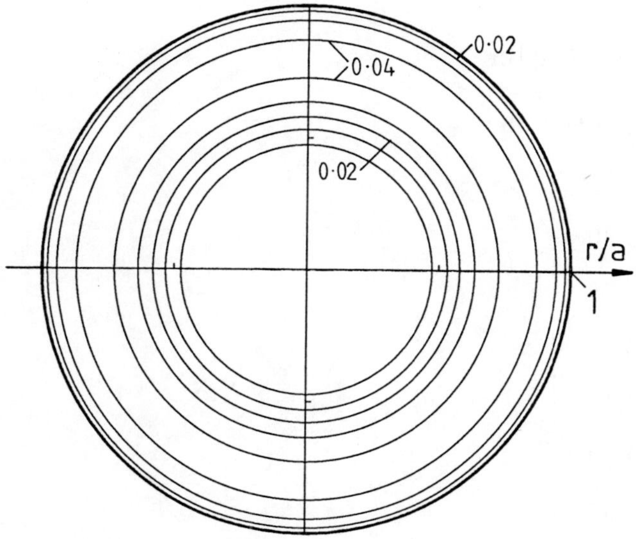

Figure 9.5: Contours of damage parameter $\tau.\delta$ over the contact patch for a Mindlin-type contact.

particular the presence of a tensile stress parallel to the surface will encourage crack formation, whereas a compressive stress will inhibit it. Ruiz and Chen recognised this and proposed a further parameter $\sigma.\tau.\delta$, obtained by multiplying $\tau.\delta$ by the maximum value of the tensile stress occurring at the same surface point in the fretted body. They found that the maximum values of this parameter occurred at locations where cracks were found experimentally. This quantity, termed the initiation parameter, may be taken to characterise the probability of crack initiation at any point in the contact.

The $\sigma.\tau.\delta$ parameter has been employed in a number of other, quite different, geometries. Nowell and Hills (1990) use it as a basis for explaining the size effect observed by Bramhall (1973) and Nowell (1988). Figure 9.6 shows the variation of $\tau.\delta$ and $\sigma.\tau.\delta$ in the trailing slip zone of the cylindrical contact employed by Nowell. The initiation parameter shows a sharp peak close to the trailing edge of the contact, where cracks usually start in practice. Both parameters fall to zero in the stick zone (where δ is zero) and outside the contact (where τ is zero). Kuno et al. (1989) applied the concept to the sphere on flat configuration and also found that cracks initiated where the initiation parameter was large. Figure 9.7 is a contour plot of the variation of $\sigma.\tau.\delta$ over the contact region for a typical case. Raoof (1993) has proposed the use of a similar parameter to characterise the degree of fretting at interwire contacts in wire ropes. Good predictions of fatigue life were obtained by this method.

The use of such damage and initiation parameters has some application in the assessment of components subjected to fretting and the prediction of likely sites for

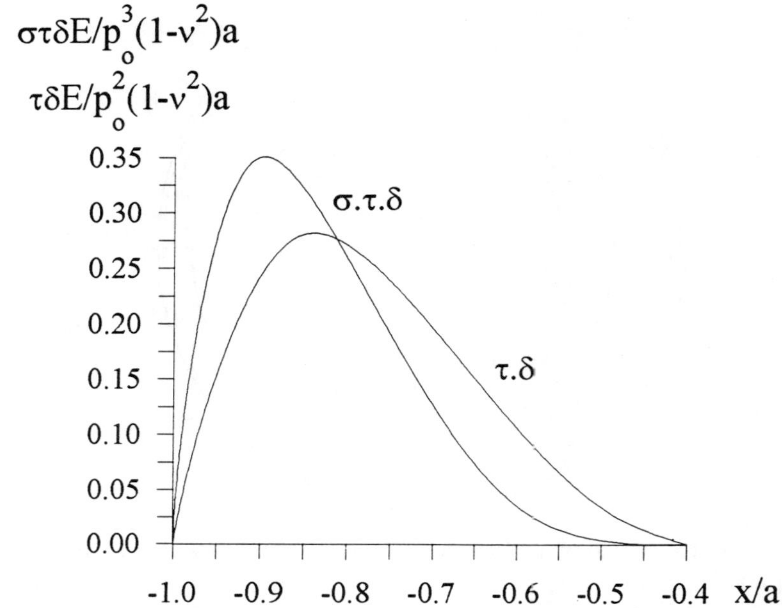

Figure 9.6: Variation of damage parameter, $\tau.\delta$, and initiation parameter, $\sigma.\tau.\delta$ in the trailing slip zone for contact of cylinders under normal and tangential load.

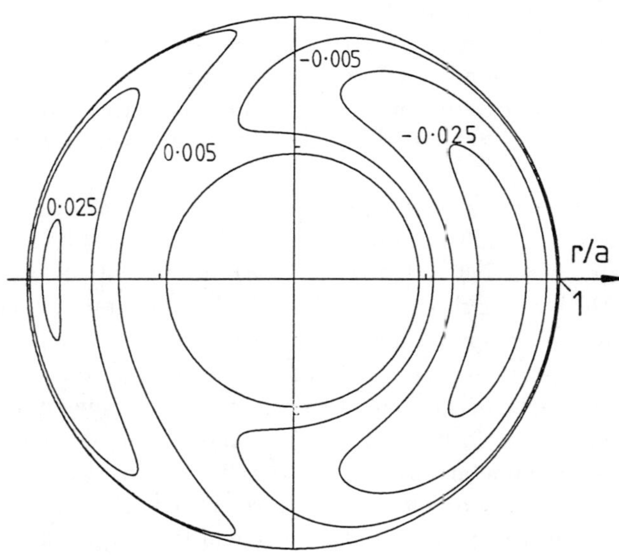

Figure 9.7: Variation of the initiation parameter, $\sigma.\tau.\delta$, over the contact region for a typical example of the sphere on flat fretting configuration.

the initiation of cracks. However, although the damage parameter, $\tau.\delta$, has a simple physical interpretation, the same cannot be said for the initiation parameter $\sigma.\tau.\delta$, although Nowell and Hills (1990) discuss a possible physical basis. There are severe limitations to its application. No account is taken of factors such as surface roughness, which might be expected to influence fretting fatigue life, other than indirectly through the influence of the friction coefficient on stresses and displacements. It is attractive to postulate that there might be some critical value for $\sigma.\tau.\delta$ at which cracks would form and that this could be regarded as some form of material constant. However it seems unlikely that this would be the case, given the multitude of other factors involved (e.g. material microstructure and environmental effects). The parameters should therefore be considered as purely empirical, but give some indication of the likely amount of damage and probability of crack initiation. They can be useful in the design of components subjected to fretting as the effects of changes in geometry can be assessed. If further understanding of crack initiation is required, a more detailed local model of the initiation process will have to be formulated and this approach will be pursued in §9.4.

One further application of the damage and initiation parameters is in investigating the effects of palliatives on crack initiation. Many different palliatives have been suggested for fretting fatigue (Beard, 1993) and often these involve the application of a surface coating. The influence of such protective coatings cannot extend much beyond the depth of the layer itself, and it is therefore apparent that the principal effects will be on crack initiation (although early phases of crack propagation may also be affected through the presence of induced residual stresses). One of the major effects of the application of a surface coating may be to change the coefficient of friction (see §5.4). In a contact in partial slip, the results of such a change on the intensity of fretting are not necessarily obvious. Under conditions of imposed dead load (as opposed to imposed displacement), an increase in the coefficient of friction will enable a greater proportion of the load to be carried in the slip zones. There will consequently be higher shear tractions in these regions, but lower displacements. The use of the $\tau.\delta$ and $\sigma.\tau.\delta$ parameters should allow the most damaging regime to be identified. Ruiz and Chen (1986) used such a procedure to explain an enhanced fretting fatigue life at high temperature by assessing the effects of a reduction in friction coefficient caused by surface oxidation.

9.4 Local modelling of crack initiation

A global model of the crack initiation process can go some way towards indicating the locations where fretting fatigue cracks are likely to initiate and to estimate the severity of fretting encountered in a particular situation. Such models cannot, however, give rise to any understanding of the initiation process and are limited in their application. No account is taken of the effects of surface condition, other than indirectly through its influence on the friction coefficient and it is difficult to incorporate material properties so as to predict the initiation life for a given experimental or practical case. For further understanding of the initiation process it is necessary to turn to more detailed local or micromechanics models which attempt to model the actual deformation mechanisms which lead to the nucleation of flaws. We cannot hope to model the exact process occurring in an individual test specimen since we would need to know precise details of grain orientation, surface profiles etc., which are impractical to measure. Instead we hope to establish a model which is typical of conditions experienced at the interface between two contacting bodies on a microscopic scale and which we can use to examine the effects of the salient variables and thus predict where initiation is likely to occur within a given macroscopic contact and how many cycles might be expected before an observable flaw is seen.

As we have seen in Chapter 5, there is always a scale at which we can consider contacts to be microscopically rough, and real contacts therefore consist of multiple asperity contacts separated by areas of zero pressure. The scale of an initiated flaw in many components will be of the same order as the asperity spacing as opposed to the scale of the overall contact. It is therefore inevitable that any micromechanics model of the initiation process will need to take into account the effects of individual asperity contacts rather than the global effects of the overall contact. The first step is to establish an appropriate model for asperity contact. In the majority of surfaces, asperities will have similar dimensions in two perpendicular directions in the plane of the surface so that an axi-symmetric model of asperity contact is more appropriate than a two-dimensional one. A useful model can be obtained by assuming the asperities to be axi-symmetric with radius of curvature ρ. If elastic behaviour is assumed, then Hertzian contacts will result and it is possible to use the results of §2.5.1 together with measurements of ρ and typical asperity spacing to predict the likely peak contact pressure occurring. If this is done, a value is often obtained for p_0 which is in excess of the yield stress of the material. At first sight this suggests that an elastic model is inappropriate but this is not necessarily the case. For purely normal Hertzian contact a near-spherical state of stress exists on the surface beneath the contact itself so that there is little tendency to yield there. Yield occurs first at a subsurface point about $0.48a$ beneath the surface on the axis of symmetry and Tresca's yield criterion predicts that yield commences at a value of p_0 given by

$$\frac{p_0}{\sigma_y} = 1.61 \tag{9.1}$$

Even after yield has commenced the zone of plastic material is completely enclosed by an elastic hinterland so that the displacements within the body (and therefore the distribution of normal traction) may not differ markedly from an extended elastic case. Elastic-plastic contact analyses are far from straightforward but a useful summary of results obtained by a number of authors is given by Johnson (1985). Significant deviation from the elastic solution does not generally occur until the normal load is at least twice that at which first yield took place. If loading proceeds beyond this some flattening of the normal traction distribution is manifested. Hardy, Baronet and Tordion (1971) found that, for the case of contact of a rigid sphere with an elastic/perfectly plastic half space, the normal pressure distribution became almost uniform (with a value of 2.3 σ_y) when the load had reached 16 times that required for first yield. An appropriate model for asperity contact will therefore depend on the surface profiles and loads encountered in any given situation. Because asperities will have different heights and peak curvatures any given contact will probably consist of a mixture of lightly deformed elastic contacts together with those exhibiting considerable plasticity. A useful guide to the amount of plasticity present when a pair of similar materials is brought into contact is given by the plasticity index Ψ, defined by (Johnson 1985)

$$\Psi = \frac{2}{AH}\sqrt{\frac{\sigma_s}{\rho}} \tag{9.2}$$

where A is the composite elastic compliance of the contacting materials (2.17), H is the hardness of the material, σ_s is the standard deviation of peak heights and ρ is a characteristic asperity radius of curvature. It will be noted that this parameter is independent of the load. This is because the number of asperities in contact increases in proportion with the applied load and therefore the mean load carried by an individual asperity remains constant[2].

9.4.1 Asperity passing model

One of the main objectives of a micromechanics model of crack initiation in fretting fatigue must be to explain the influence of slip amplitude on fatigue life. As discussed earlier, it is clear that slip amplitude cannot affect the propagation phase of crack life, which is controlled purely by the stress state (for a given set of environmental and material parameters). Once it is realised that interaction takes place between the surfaces on an asperity scale it is relatively straightforward to propose an explanation based on the number of asperity passes experienced in any one cycle. Consider the situation shown in fig. 9.8. During part of the fretting cycle there will be relative slip between the pad and the specimen. A particle, A, on the surface of the fatigue specimen will experience the surface of the pad moving past it. If the slip amplitude is large enough a number of asperities will pass over point A before the slip direction reverses. Each asperity contact will have an associated subsurface stress field which will move with the asperity. If the asperities are relatively widely spaced, particle A will no longer be experiencing significant stresses from an asperity which has passed when the next arrives. Thus the stresses experienced by surface (and immediately subsurface) particles will vary cyclically, with one cycle for each **asperity pass**. There will, of course be an additional stress cycle due to any bulk load in the specimen and this will be at the main loading frequency rather than that of asperity passing. The

[2]For further discussion of this point see Johnson (1985) or Hills, Nowell, and Sackfield (1993).

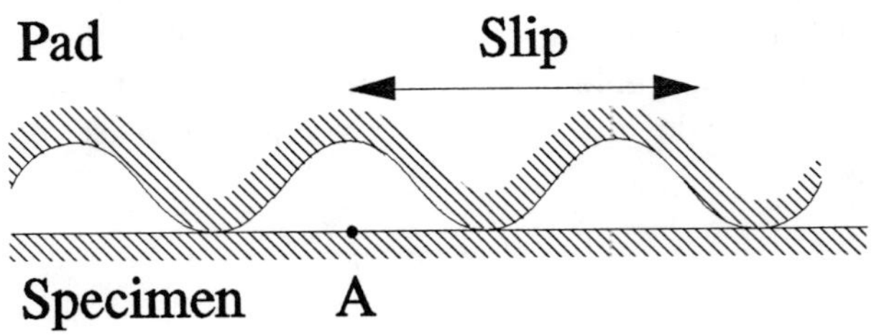

Figure 9.8: Micromechanics model of relative motion in a slip zone during fretting contact.

overall loading cycle experienced will be as shown in fig. 9.9. The explanation for the decrease of initiation life with increased slip amplitude is now apparent. The larger the slip amplitude, the more asperity cycles are accumulated for each major cycle of the load and the faster the fatigue damage accumulates to nucleate a flaw. It will be seen from fig. 9.9 that asperity passes experienced when the bulk load is high (termed forward slip in the figure) will be more damaging than those at the low bulk stresses experienced in reverse slip. Bower and Johnson (1989) show that accumulation of plastic strain is not possible if a contact load alone is passed backwards and forwards across a surface. In the case of fretting with a bulk load, however, the effect of the bulk stress is to render asperity passes in one direction more severe than those in the other and hence accumulation of strain is possible.

The asperity-passing model does not explain the mechanisms of damage experienced in the specimen but it does go some way towards explaining the effectiveness of the empirical bulk parameters $\tau.\delta$ and $\sigma.\tau.\delta$. As remarked above, for a given pair of contacting rough surfaces the number of contacting asperities is proportional to the load. Hence a measure of the density of asperity contacts can be obtained from the mean local contact pressure p (as predicted by a model of the equivalent smooth contact). Multiplying this by the coefficient of friction to obtain the shear traction τ includes a measure of the severity of each asperity stress cycle since stresses due to a sliding asperity will be more severe if the coefficient of friction is higher. The amplitude of relative slip, δ, gives a measure of the number of asperity passes in one overall cycle of the fretting load. Hence $\tau.\delta$ is a good measure of the damage caused by asperity passes in a single fretting cycle.

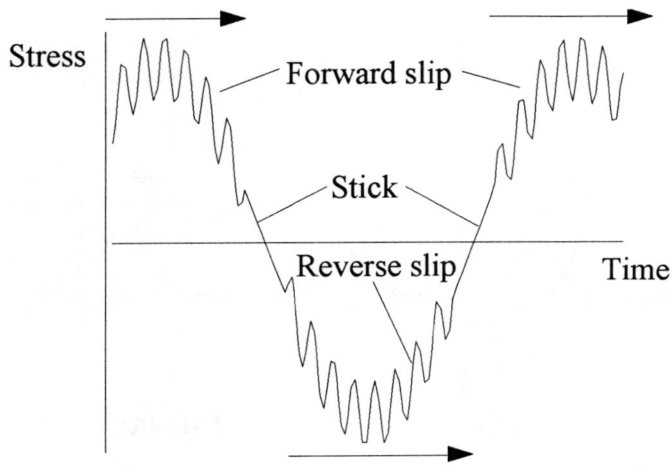

Figure 9.9: Stress cycle experienced by near-surface points in the specimen as a result of combined bulk and asperity loading.

The inclusion of the in-plane tensile stress, σ, to form the initiation parameter $\sigma.\tau.\delta$ is less easy to justify physically and it is necessary to fall back on the argument that both fretting damage and a tensile stress are necessary to initiate cracks and hence the product $\sigma.\tau.\delta$ is a good measure of this joint requirement.

One of the important features of the asperity-passing model is that it highlights a number of important length scales in the initiation process. These are, first, those dimensions associated with the geometry of the component:

(i) the amplitude of micro-slip,

(ii) the asperity spacing, and

(iii) the size of an average asperity contact patch.

As has been shown, providing that dimension (iii) is much less than (ii), which will often be the case, the number of load reversals 'seen' by surface particles will be geared up from the basic number of bulk load reversals by the ratio of dimension (ii)/dimension (i). This gearing ratio will persist only while the embryo crack is comparable in length with the asperity size. Thus, the accelerated cycling which is essential to the nucleation process will persist over a characteristic dimension. A second length scale which may now be introduced is that of the grain size. It is acknowledged that if a nucleating crack spans several grains it exists in something approaching a continuum, and therefore fracture mechanics ideas become appropriate. Thus, if the grain size prevalent in the material is small compared with the asperity pitch, 'gearing' will be present during much of the nucleation process. If, on the other hand, the material has a very coarse-grained structure, it is quite possible to conceive of cases where the crack still exists within a single grain, although the crack

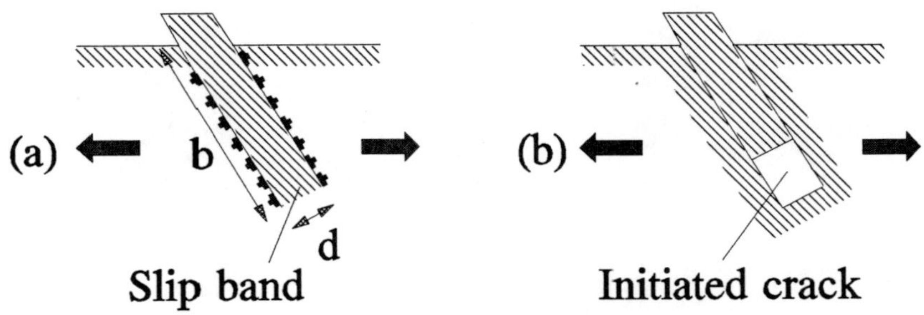

Figure 9.10: (a) Formation of an extrusion on the surface of a material by slip along persistent slip bands, and (b) an initiated crack formed when the accumulated slip is released.

tip has grown out of the zone where gearing accelerates growth. Thirdly, the other characteristic dimension which plays an important part is the macroscopic size of the contact itself. This determines the scale of the contact stress field, and hence the characteristics of the stress field in the neighbourhood of the crack tip when it passes from stage 1 to stage 2 growth, and the crack trajectory changes.

9.4.2 Persistent slip band model

Further progress with a micromechanics model of crack initiation demands a consideration of the details of the crack initiation process itself in addition to the asperity contact which is one of the main driving forces. There is a considerable body of experimental evidence that crack initiation takes place by slip along persistent slip bands. This leads to the formation of extrusions and intrusions on the surface of the material (fig. 9.10) which then develop into fatigue cracks. An excellent summary of much of the experimental work can be found in Suresh (1991). The slip is driven by shear stresses and is found to be partly irreversible (a friction stress must be overcome before reverse slip can occur). Forward slip takes place along one plane during the loading part of the fatigue cycle. When unloading commences, reverse slip takes place preferentially along an adjacent slip plane and an extrusion is formed. Recent work by Venkataraman et al. (1989) and Mura and Nakasone (1990) has resulted in a model for this initiation process. The slip is modelled by arrays of dislocations in much the same way as the modelling of cracks was undertaken in Chapter 6. Two parallel slip planes are installed which are taken to pile up against a subsurface obstacle such as an inclusion or grain boundary (fig. 9.10a). The slip planes are subjected to cycles of load and the distribution of dislocations calculated so that the combined influence of applied load together with the interaction stress due to the presence of the other dislocations is equal to the friction stress at each point on either slip plane, i.e.

$$\tau_1^D + \tau_2^D + \tau^A = \pm\tau_f \tag{9.3}$$

where τ_i^D is the stress caused by dislocations on layer i, τ^A is the applied shear stress, and τ_f is the friction stress. The right hand side is taken to be positive on one slip plane and negative on the other. As the slip planes are subjected to cycles of loading it is found that slip accumulates on each slip plane and an extrusion forms. The stored strain energy is calculated after each loading cycle. At some point it becomes energetically favourable for an embryo crack to form (fig. 9.10b). The length of this crack may be calculated by assuming that the locked-in slip is completely released. By evaluating the energy change if the crack were to form it is possible to determine when crack initiation becomes preferable and hence the number of loading cycles to initiation may be calculated. There is clearly a large number of variables in this analysis, including materials parameters such as the friction stress, τ_f, and the surface energy of the crack faces formed, γ_s, together with geometrical variables such as the length of the initial slip planes, c, the separation of the slip planes, d, and the slip band inclination, θ. This means that it is unreasonable to expect to be able to calculate the number of initiation cycles in any particular configuration, because many of the variables will be unknown, or difficult to determine. However it is possible to examine the effect of these variables on the initiation life.

It is now possible to extend this model, which was established for plain fatigue, to address crack initiation under the influence of asperity loading. In the slip zone of a (macroscopic) contact the slip amplitudes caused by tangential loading will be greater than that which can be accommodated in the compliance of individual asperities. Hence such asperities will slide during part of the loading cycle. It is therefore appropriate to begin by examining the shear stresses produced by a sliding asperity. It is these stresses which we anticipate will drive crack initiation by a slip process similar to that modelled above. The asperity is taken as a Hertzian contact between a sphere and a plane and subject to Coulomb friction during sliding. The results of §2.5.1 may be employed to calculate the resulting stress field. Figure 9.11 shows the shear stresses occurring on a plane at 45° to the surface for a point over which the asperity is sliding backwards and forwards. It will be seen that, whilst the point is within the contact patch, the shear stress on this planevaries linearly with asperity position. Once the edge of the asperity passes over the point of interest the stresses fall away rapidly. This leads to another possible explanation of the variation of initiation life with slip amplitude. If the amplitude of asperity sliding is less than a', the stress cycle produced will be similar to that marked ABCD in fig. 9.11. The amplitude of shear stresses experienced on the 45° plane will therefore increase with slip

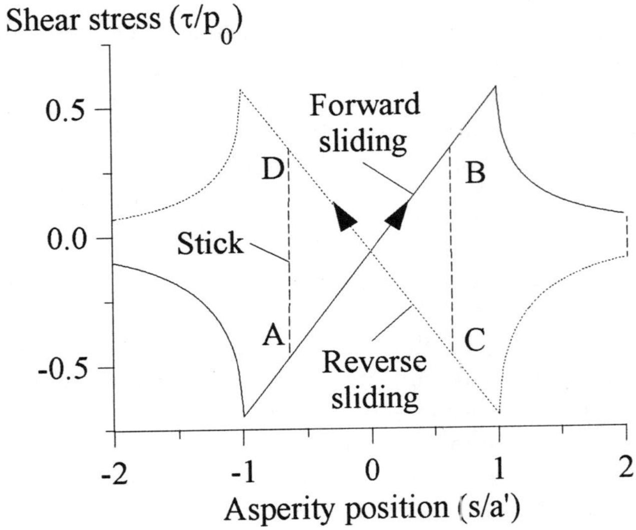

Figure 9.11: Shear stresses on a plane at 45° to the surface for a point on the surface beneath a sliding asperity. Here p_0 is the peak pressure of the asperity contact, a' the aperity contact radius, $f = 0.75$ and $\nu = 0.3$.

amplitude. The maximum amplitude will occur when the asperity passes completely over the surface point; further increases in slip will not produce a more damaging stress cycle until the slip amplitude is sufficiently large to bring a second asperity into the neighbourhood of the surface point.

To apply the shear stresses produced by a sliding asperity to a the slip band model the configuration depicted in fig. 9.12 is considered. Let us include only the effect of a single asperity and ignore, for the time being at least, loading from any bulk stress in the specimen or other nearby asperities. If we are to obtain any variation of initiation life with amplitude of microslip, it is necessary to assume that the asperity passes over the slip bands during at least part of the cycle. Unless this is the case the amplitude of the shear stress experienced is controlled purely by the closest distance between asperity and slip band (see fig. 9.11). Since the asperity passes over the slip band, we must assume that any extruded material is smeared out over the surface by the passing contact.

We can now formulate the model using precisely the method of Mura and Nakasone (1990), save that the stresses driving dislocation slip are those caused by the sliding asperity. Further details of the technique are given by Nowell, Hills, and Dai (1993). Figure 9.13 shows some sample results obtained from this model. Once again, several important length scales emerge. These are, the amplitude of slip, the asperity contact width and the length of the slip bands. Typically it might be expected that slip would pile up against a grain boundary and hence the length of the slip bands might be equated with the grain size of the material. If the slip amplitude is large

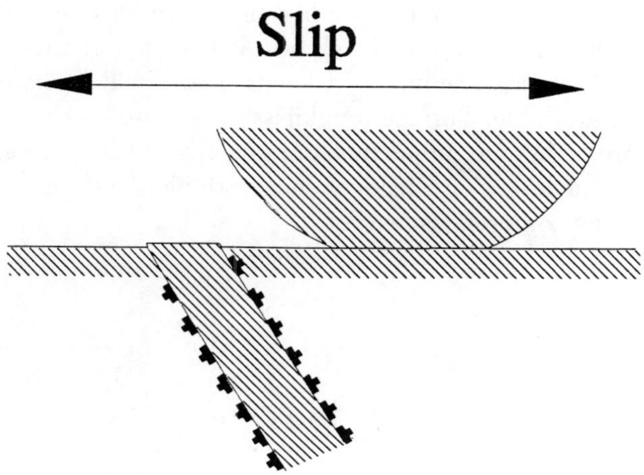

Figure 9.12: Slip-band model for crack initiation in fretting fatigue under asperity loading.

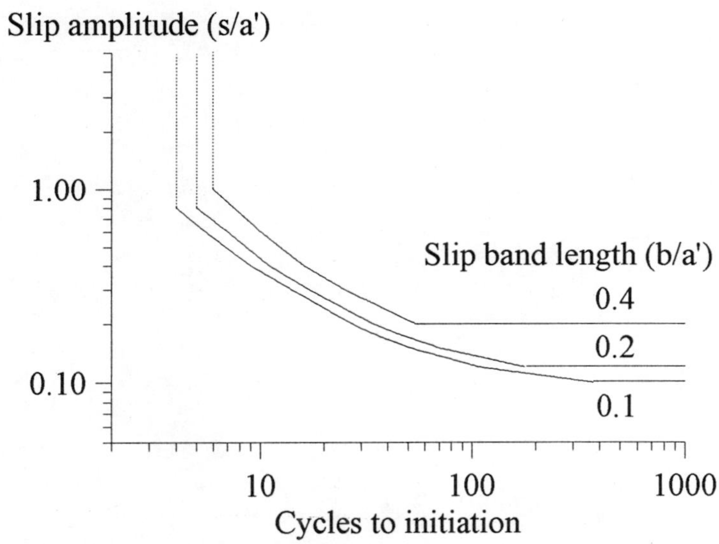

Figure 9.13: Variation of initiation life with asperity slip amplitude for a sample case as predicted by a slip-band model. Dimensions are normalised with respect to a', the semi-width of the asperity contact patch.

enough to cause multiple asperity passes then the asperity spacing again becomes an important dimension. This type of slip-band initiation model is still in its infancy, particularly when applied to fretting fatigue, but the approach has a number of clear advantages. The dependence of initiation life on slip amplitude can be explained in a credible way without resorting to 'black-box' arguments of frictional energy density. Moreover it highlights a number of length scales which should be important in determining initiation life. Surface roughness parameters such as asperity density and peak curvature emerge as important parameters which we might expect to influence fretting fatigue life. The effects of such variables have yet to be investigated experimentally in any comprehensive way.

Chapter 10

Conclusions

In a study of a field such as fatigue, which depends at least partly on an ever more refined understanding of materials phenomena, there can be no final statement completely explaining the physics of the problem. In the case of some kinds of failure, such as yield, or brittle fracture of a flawed component, it is, nevertheless, possible to make simple concise statements which are of immediate usefulness to the engineering designer. In attempting to quantify fretting fatigue, even this is not possible, principally because of our incomplete understanding of the nature of crack initiation, and partly because the process is a stochastic one, depending on the vagaries of the material crystallography, surface topography and cleanliness. We should remind ourselves that the life of a crack is conveniently categorised as a two-stage process; first the crack nucleates, usually from the surface and from pre-existing defect, no matter how small that defect may be. It is this phase of the life which is so critically dependent on the quantities cited, and which is difficult to quantify. Secondly, the crack propagates, in a direction and at a rate controlled also by the state of stress, but this phase is somewhat less sensitive to the material properties obtaining within the individual grains. There is a considerable body of experimental evidence to show that fretting influences both of these processes. The goal of an engineering approach to the analysis of fretting fatigue is to quantify the effects, so as to allow the designer to estimate component life reliably using the results of basic materials tests.

The question we set out to answer is essentially; 'When will a fretting fatigue crack initiate, and if it does so, at what rate will it grow?'. This has no simple answer, precisely because of the difficulties in quantifying initiation. Our understanding of the problem indicates that a knowledge of the states of stress, strain, and displacement existing on a macroscopic scale can *never* provide a completely satisfactory prediction of the likelihood of crack initiation. Further progress will depend on improved models of initiation in general, and in this respect the fretting fatigue problem has many parallels with that of plain fatigue. The location of the point of initiation is much better defined, because of the stress concentration provided by the contact, but the stress state is far more complicated and the effects of surface interactions need to be taken into account.

Given this rather pessimistic start to our concluding remarks, it is perhaps appropriate now to turn to the more positive achievements of applied mechanics in developing an understanding of fretting fatigue. The propagation phase of the life of any crack is controlled by the stress intensity factors, and these in turn are controlled solely by the geometry of the crack and the stress field. Provided that the stress intensity factors may be found reliably, the crack growth rate may itself be accurately predicted[1]. In these respects fretting fatigue is no different from plain fatigue. The propagation phase will be predictable if reliable stress intensity factors can be obtained from a knowledge of the overall geometry and loading. It is, of course, important to recognise that, in fretting fatigue, cracks which are comparable in length with the size of contact are present in an extremely steep stress gradient, and therefore demand a rigorous means of determining the crack tip stress intensity factors. In addition it may be necessary to include the effects of residual stresses, crack face closure, and abrupt changes of crack direction on stress intensity factor calculations.

Analysis of the propagation phase of fretting fatigue has two main applications: in the interpretation of experimental results, and in the prediction of component life in practical situations (such as the flange first shown in fig. 1.1). In the latter case the stresses must generally be found by numerical techniques, but there is a range of simple geometrical configurations, described earlier, for which the state of stress is very well defined and which permit closed form solutions to be found for all the field variables. These configurations are particularly attractive for fretting experiments, since many of the numerical difficulties which might be present in more complicated geometries are removed. A study of the simpler geometries described in this book has two further useful purposes: first it indicates the complex nature of contacts existing under partial slip conditions and highlights the inequalities which must be observed in order fully to specify the stick-slip regime present. Without the understanding developed from the study of these contacts it is extremely difficult to formulate numerical models of contact problems and to interpret the results obtained. Secondly, simple contact models provide a means of analysing the contact at an asperity scale, albeit under conditions of fully elastic behaviour.

Returning to the question of the quantification of initiation, two approaches have been suggested; one based on bulk field variables (frictional energy expenditure density and in-plane surface tension), and the other on the micromechanics of the formation of extrusions. The first has the obvious merit that it needs only a standard engineering knowledge of the state of stress, friction etc., but lacks rigour in taking physical phenomena into account. The second develops a representative model of the initiation process, but demands a knowledge of physical quantities such as the friction stress for a dislocation which are not normally measured in standard engi-

[1]It should be noted that the exponent, m, in the Paris equation (6.13) is usually about 4, so that a 10% error in determining the stress intensity factor will lead to an error of approximately 40% in the predicted crack propagation rate.

neering applications. It remains to be seen whether a simple correlation of the field variables will be adequate for design purposes; even then the experiments must be conducted under very similar conditions to those occurring in the prototype in order to establish, for example, critical values of $\tau.\delta$. This approach would seem to have attractions as a design tool, but micromechanics models must be developed for research use and to validate a 'bulk parameter' approach. Although it is impossible, at this stage, to quantify our results by stating criteria for 'no crack growth', or, indeed, predicting the number of cycles to failure the analysis does provide valuable insight. Our understanding of the load transfer characteristics between components is improved by elucidating concepts of partial slip; the influence of modifying surface friction, either by dint of employing palliatives, or by changing the surface roughness, has been explained. This permits identification of areas of weakness in any design, to which more detailed attention may be devoted, both analytically and experimentally.

An implicit assumption made at several points in the text is that the macroscopic geometry of the contact does not enter into the problem except insofar as it defines the contact pressure distribution and other field variables, i.e. we assume that the problems of debris entrapment and the stress concentrations this is known to produce are not pertinent. Within the confines of our own experimental work this would seem to be borne out, as fretting wear did not play an important rôle. The assumption that fretting effects are geometry-independent leads us to the assertion that fretting fatigue tests are often best carried out not with a simulation of an actual component contact, which may have limited applicability, but under very carefully controlled conditions both of the field variables and the surface finish. In turn this suggests the use of simple convex contacts within an experiment, as this will reduce the sensitivity of the experiment to manufacturing and alignment problems. Further, the physical influence of other properties of the system such as the grain size and morphology, together with the aggressiveness of the environment (in particular its corrosive qualities) must be maintained constant from test to test for the results to be carried over. A reliable knowledge of the state of stress is especially helpful in predicting the probable locations of crack initiation. Further, the procedures described for determining crack-tip stress intensity factors enable their values to be determined efficiently and with precision, even in cases where the crack flanks are pressed together along part of their length, or the crack is kinked. It is only when the experimental configuration adopted is well understood, and the conditions are repeatable, that reliable materials data will be obtained. Such data are a necessary pre-requisite for the prediction of fatigue life in real components. One of the principal conclusions of our work must be that the adoption of well defined experimental configurations based on the Hertzian contact is essential if test results are to be useful in practice.

We hope that this book will both provide useful design information on the avoidance of fretting fatigue problems, and to stimulate researchers in the field into examining problems in a different light. In particular, we wish to emphasise the key rôle applied mechanics has to play in analysing fretting, and the care with which test geometries need to be chosen in order to produce both meaningful and widely applicable results. We have presented solutions for many idealised contact problems, and for crack tip stress intensity factors relevant to cracks propelled by a contact stress field. Their descriptions contained within this book will enable the reader to reproduce reliable solutions needed for the analysis of experiments. However, for those wishing to use existing tested software, a range of programs is available on application to the authors. The programs available are supplied as FORTRAN subroutines so that the user may readily incorporate them into his or her own software. Details of the available programs, which include routines for the evaluation of stresses and stress intensity factors may be obtained by writing to:

Dr D.A. Hills (contact mechanics software),
c/o Lincoln College,
OXFORD, OX1 3DR, U.K.

Appendix A

Kernels for a dislocation in a half-plane

As used in equation (6.37)

$$
\begin{aligned}
G_{xxx} &= y\left\{\frac{-2}{r_1^2} - \frac{4(x-c)^2}{r_1^4} + \frac{2}{r_2^2} + \frac{4(x+c)^2}{r_2^4} - \frac{8c(x+c)}{r_2^4} + \frac{32c(x+c)^3}{r_2^6}\right. \\
&\qquad \left. + \frac{8c^2}{r_2^4} - \frac{32c^2(x+c)^2}{r_2^6}\right\}
\end{aligned}
$$

$$
\begin{aligned}
G_{xyy} &= y\left\{\frac{-2}{r_1^2} + \frac{4(x-c)^2}{r_1^4} + \frac{2}{r_2^2} - \frac{4(x+c)^2}{r_2^4} + \frac{24c(x+c)}{r_2^4} - \frac{32c(x+c)^3}{r_2^6}\right. \\
&\qquad \left. - \frac{8c^2}{r_2^4} + \frac{32c^2(x+c)^2}{r_2^6}\right\}
\end{aligned}
$$

$$
\begin{aligned}
G_{xxy} &= \left\{\frac{-2(x-c)}{r_1^2} + \frac{4(x-c)^3}{r_1^4} + \frac{2(x+c)}{r_2^2} - \frac{4(x+c)^3}{r_2^4} - \frac{4c}{r_2^2} + \frac{32c(x+c)^2}{r_2^4}\right. \\
&\qquad \left. - \frac{32c(x+c)^4}{r_2^6} - \frac{24c^2(x+c)}{r_2^4} + \frac{32c^2(x+c)^3}{r_2^6}\right\}
\end{aligned}
$$

$$
\begin{aligned}
G_{yxx} &= \left\{\frac{-2(x-c)}{r_1^2} + \frac{4(x-c)^3}{r_1^4} + \frac{2(x+c)}{r_2^2} - \frac{4(x+c)^3}{r_2^4} - \frac{4c}{r_2^2} - \frac{16c(x+c)^2}{r_2^4}\right. \\
&\qquad \left. + \frac{32c(x+c)^4}{r_2^6} + \frac{24c^2(x+c)}{r_2^4} - \frac{32c^2(x+c)^3}{r_2^6}\right\}
\end{aligned}
$$

$$
\begin{aligned}
G_{yyy} &= \left\{\frac{6(x-c)}{r_1^2} - \frac{4(x-c)^3}{r_1^4} - \frac{6(x+c)}{r_2^2} + \frac{4(x+c)^3}{r_2^4} - \frac{4c}{r_2^2} + \frac{32c(x+c)^2}{r_2^4}\right. \\
&\qquad \left. - \frac{32c(x+c)^4}{r_2^6} - \frac{24c^2(x+c)}{r_2^4} + \frac{32c^2(x+c)^3}{r_2^6}\right\}
\end{aligned}
$$

$$G_{yxy} = y\left\{\frac{-2}{r_1^2} + \frac{4(x-c)^2}{r_1^4} + \frac{2}{r_2^2} - \frac{4(x+c)^2}{r_2^4} - \frac{8c(x+c)}{r_2^4} + \frac{32c(x+c)^3}{r_2^6} + \frac{8c^2}{r_2^4} - \frac{32c^2(x+c)^2}{r_2^6}\right\}$$

where

$$r_1^2 = (x-c)^2 + y^2$$
$$r_2^2 = (x+c)^2 + y^2$$

References

Pages on which references occur are indicated by the number in square brackets.

Abramowitz, M., and Stegin, I. (1972) *Handbook of Mathematical Functions, Second Edition*, Dover, New York. [85]

Aliabadi, M.H., and Brebbia, C.A. (1993) *Computational Methods in Contact Mechanics,* Computational Mechanics Publications, Southampton. [89]

Aliabadi, M.H., and Rooke, D.P. (1991) *Numerical Fracture Mechanics*, Kluwer, Dordrecht, Netherlands. [136]

Aliabadi, M.H., Rooke, D.P., and Cartwright, D.J. (1987) Mixed mode Bueckner weight functions using boundary element analysis, *Int. Jnl. Fract.*, **34**, 131-147. [136]

Amontons, G. (1699) De la resistance cause dans les machines, *Memoires de L'Acedemie Royale*, A, 275-282. [41]

Arnell, R.D., Davies, P.B., Halling, J., and Whomes, T.L. (1991) *Tribology: Principles and Design Applications*, Macmillan, London. [115]

Attia, M.H, and Waterhouse, R.B. (1992) *ASTM STP 1159: Standardization of Fretting Fatigue Test Methods and Equipment,* ASTM, Philadelphia. [3]

Barber, J.R. (1983) The solution of elasticity problems for the half-space by the method of Green and Collins, *Appl. Sci. Res.*, **40**, 135-157. [28, 31]

Barber, J.R. (1992) *Elasticity*, Kluwer, Dordrecht. [98, 157]

Beard, J. (1993) Palliatives for fretting fatigue, *Proc. International Conf. on Fretting Fatigue, Sheffield, April 1993*, I.Mech.E., London. [205]

Becker, A.A. (1992) *The Boundary Element Method in Engineering*, McGraw-Hill, London. [89]

Bentall, R.H., and Johnson, K.L. (1968) An elastic strip in plane rolling contact, *Int. Jnl. Mech. Sci.*, **10**, 637-663. [91]

Benthem, J.P. (1977) State of stress at the vertex of a quarter-infinite crack in a half-space, *Int. Jnl. Solids Structures*, **13**, 479-492. [184]

Bogy, D.B. (1968) Edge-bonded dissimilar orthogonal elastic wedges under normal and shear loading, *Jnl. Appl. Mech.*, **35**, 460-466. [159]

Bogy, D.B. (1971) Two edge-bonded elastic wedges of different materials and wedge angles under surface tractions, *Jnl. Appl. Mech.*, **38**, 377-386. [159]

Bower, A.F., and Johnson, K.L. (1989) The influence of strain hardening on cumulative plastic deformation in rolling and sliding line contact, *Jnl. Mech. Phys. Solids*, **37**, 4, 471-494. [208]

Bramhall, R. (1973) *Studies in fretting fatigue* D.Phil. thesis, University of Oxford. [4, 162, 164, 189, 192, 199, 203]

Broek, D. (1982) *Elementary Engineering Fracture Mechanics*, Nijhoff, The Hague. [1, 127]

Broek, D. (1989) *The Practical Use of Fracture Mechanics*, Kluwer, Dordrecht. [1, 127]

Bueckner, H.F. (1958) The Propagation of Cracks and the Energy of Elastic Deformation. *Trans. ASME*, **80**, 1225-1230. [135]

Cattaneo, C. (1938) Sul contatto di due corpi elastici: distribuzion locale degli sforzi, *Reconditi dell Accademia nazionale dei Lincei*, **27** , 342-348, 434-436,474-478. [50]

Chang, F-K, Comninou, M. and Barber, J.R. (1983) Slip Between a Layer and a Substrate Caused by a Normal Force Moving Steadily over the Surface, *Int. Jnl. Mech. Sci.*, **25**, 803-809. [147]

Chivers, T.C., and Gordelier, S.C. (1985) Fretting fatigue and contact conditions : a rational explanation of palliative behaviour, *Proc. I. Mech.E.*, **199**, C4, 325-337. [125]

Collins, W.D. (1959) On the solution of some axisymmetric boundary value problems by means of integral equations, II: Further problems for a circular disc and a spherical cap, *Mathematika*, **6**, 120-133. [28]

Comninou, M. (1976) Stress singularities at a sharp edge in contact problems with friction, *ZAMP*, **27**, 493-499. [160]

Comninou, M. (1977) The Interface Crack, *Jnl. Appl. Mech*, **44**, 631-636. [148]

Comninou, M., and Barber, J.R. (1983) Frictional slip between a layer and a substrate due to a periodic tangential surface force, *Int. Jnl. Solids Structures*, **19**, 6, 533-539. [119]

Comninou, M., Barber, J.R. and Dundurs, J. (1983) Interface slip caused by a surface load moving at a constant speed, *Int. Jnl. Mech. Sci.*, **25**, 41- 46. [147]

Coulomb, C.A. (1785) Theorie des machines simples, *Memoires de Mathmatique et de Physique de l'Acedmie Royale*, 161-342. [41]

Crouch, S.L., and Starfield, A.M. (1983) *Boundary Element Methods in Solid Mechanics*, Unwin Hyman, London. [89]

Dai, D.N., Nowell, D., and Hills, D.A. (1993) Eigenstrain methods in 3-D crack problems: an alternative integration procedure, *Jnl. Mech. Phys. Solids*, **41**, 6, 1003-1017. [152, 184]

Derby, B., Hills, D.A., and Ruiz, C. (1992) *Materials for Engineering: A Fundamental Approach.* Longman Scientific and Technical, Harlow, Essex. [127, 129, 131, 149, 171]

Deresiewicz, H. (1954) Contact of elastic spheres under an oscillating torsional couple, *Jnl. Appl. Mech.*, **21**, 52-56. [80, 81, 82]

Dewynne, J.N., Hills, D.A. and Nowell, D. (1992) The Opening Displacement of Surface-Breaking Plane Cracks. *Comp. Meth. in Appl. Mech. and Eng.*, **97**, 321-331. [139]

Dieter, G. (1961) *Mechanical Metallurgy.* McGraw-Hill, Tokyo. [127, 130]

Dubourg, M.C. and Colin, F. (1992) On propagation of multiple cracks' *Proc. Contact Mechanics International Symposium, Lausanne, Oct. 1992,* Ed. A. Curnier, Presses Polytechniques et Universitaires Romandes, p507. [193]

Dugdale, D.S. (1960) Yielding of steel sheets containing slits, *Jnl. Mech. Phys. Solids*, **8**, 100-108. [174]

Dundurs, J. (1969) Discussion of edge-bonded dissimilar orthogonal elastic wedges under normal and shear loading, *Jnl Appl. Mech.*, **36**, 650-652. [17]

Dundurs, J. (1975) Properties of elastic bodies in contact, in *Mechanics of Contact between Deformable Bodies,* Ed. de Pater and Kalker, Delft Univ. Press, Delft. [12]

Dundurs, J., and Lee, M-S (1972) Stress concentration at a sharp edge in contact problems, *Jnl. of Elasticity*, **2**, 2, 109-112. [159]

Dundurs, J., and Mura, T. (1964) Interaction Between an Edge Dislocation and a Circular Inclusion, *Jnl. Mech. Phys. Solids*, **12** , 177-189. [137, 140]

Dundurs, J., and Sendeckyj, G.P. (1965) Behaviour of an edge dislocation near a bimetallic interface, *Jnl. Appl. Phys.*, **36**, 3353-3354. [137]

Erdogan, F., Gupta, G. D., and Cook, T. S. (1973) Numerical solution of singular integral equations, pp368-425 of *Methods of Analysis and Solutions of Crack Problems*, Ed. G. C. Sih, Noordhoff, Leyden . [63, 71, 83, 84, 85, 139, 143]

Erdogan, F. and Sih, G.C. (1963) On the crack extension in plates under plane loading and transverse shear, *Jnl. of Basic Engineering*, **85**, 519-527. [174]

Ewalds, H.L., and Wanhill, R.J.H. (1984) *Fracture Mechanics*, Arnold, London, and Delfste Uitgevers Maatschappij, Delft. [1, 127, 129, 136, 149]

Fan, H., Keer, L.M., and Mura, T. (1991) The effect of plastic deformation on crack initiation in fatigue, *Int. Jnl. Solids Struct.*, **28**, 9, 1095-1104. [123, 131]

Fenner, A.J., and Field, J.E. (1958) La fatigue dans les conditions de frottement, *Rev. Met.*, **55**, 475-485. [3]

Fessler, H., and Ollerton, E. (1957) Contact stresses in toroids under radial loads, *Brit. Jnl. Appl. Phys.*, **8**, 387-393. [12]

Forsyth, P.J.E. (1961) A two-stage process of fatigue crack growth, *Proc. Crack Propagation Symposium. Cranfield College of Aeronautics,* pp. 76-94. [130, 170]

Gdoutos, E.E. (1990) *Fracture Mechanics Criteria and Applications*, Kluwer, Dordrecht. [1, 127]

Gdoutos, E.E. (1993) *Fracture Mechanics, an Introduction*, Kluwer, Dordrecht. [1, 127]

Gladwell, G.M.L. (1976) On some unbonded contact problems in plane elasticity theory, *Jnl. Appl. Mech.*, **43**, 263-267. [12]

Gladwell, G.M.L. (1980) *Contact Problems in the Classical Theory of Elasticity*, Sijthoff and Noordhoff, Alphen aan den Rijn. [vi, 10, 18, 28, 85]

Goodman, L.E. (1962) Contact stress analysis of normally loaded rough spheres, *Jnl. Appl. Mech.*, **29**, 3, 515-522. [66]

Goodman, L.E., and Keer, L.M. (1965) The contact stress problem for an elastic sphere indenting an elastic cavity, *Int. Jnl. Solids Structures,* **1**, 407-415. [12]

Gordelier, S.C., and Chivers, T.C. (1979) Literature review of palliatives for fretting fatigue, *Wear*, **56**, 177-190. [122]

Green, A.E., and Zerna, W. (1968) *Theoretical Elasticity,* Second Edition, Oxford University Press, Oxford. [28]

Greenwood, J.A., and Tripp, J.H. (1967) The elastic contact of rough spheres, *Jnl. Appl. Mech.*, **34**, 153-159. [107]

Greenwood, J.A., and Williamson, J.B.P. (1966) Contact of nominally flat surfaces, *Proc. Roy. Soc. Ser. A.*, **295**, 300-319. [115]

Hamilton, G.M. (1983) Explicit equations for the stresses beneath a sliding spherical contact, *Proc. I. Mech.E.*, **197C**, 53-59. [33]

Hamilton, G.M., and Goodman, L.E. (1966) The stress field created by a circular sliding contact, *Jnl. Appl. Mech.*, **33,** 371-376. [33, 52]

Hardy, C., Baronet, C.N., and Tordion, G.V. (1971) Elastoplastic indentation of a half space by a rigid sphere, *Jnl. Num. Methods in Engineering*, **3**, 451-. [207]

Hattori, T., Nakamura, M., Sakata, H., and Watanabe, T. (1988) Fretting fatigue analysis using fracture mechanics, *JSME Int. Jnl.*, Series 1, **31**, 1, 100-107. [89]

Hills, D.A., and Comninou, M. (1985 (a)) A Normally Loaded Half-Plane with an Edge Crack, *Int. Jnl. Solids Structures*, **21**, 4, 399-410. [147]

Hills, D.A., and Comninou, M. (1985 (b)) An Analysis of Fretting Fatigue Cracks During Loading Phase, *Int. Jnl. Solids Structures*, 21, **7**, 721-730. [147]

Hills, D.A., and Nowell, D. (1989) Stress intensity calibrations for closed cracks, *Jnl. Strain Anal.,* **24**, 1, 37-43. [145, 147]

Hills, D.A., and Nowell, D. (1990) Kinked cracks: finding stress intensity factors, pp 36-50 of *Applied Stress Analysis*, Proc. Conf. on Applied Stress Analysis, Nottingham, August 1990, Ed T.H. Hyde and E. Ollerton, Elsevier, London. [182]

Hills, D.A., and Nowell, D. (1993) Discussion on 'Crack line and edge Green's function for stress intensity factors of inclined edge cracks' by D.P. Rooke, D.B. Rayaprolu, and M.H. Aliabadi, *Fatigue Fract. Engng Mater. Struct.*, **16**, 2, 267-270. [178]

Hills, D.A., Nowell, D., and O'Connor, J.J. (1988) On the mechanics of fretting fatigue, *Wear*, **125**, 129-156. [60, 189, 192]

Hills, D.A., Nowell, D., and Sackfield, A. (1988) Surface fatigue considerations in fretting, in *Interface Dynamics, Proc 14th Leeds-Lyon Symposium on Tribology,* Ed D. Dowson, C.M. Taylor, M. Godet, D. Berthe, Elsevier, Amsterdam. [117]

Hills, D.A., Nowell, D., and Sackfield, A. (1993) *Mechanics of Elastic Contacts,* Butterworth-Heinemann, Oxford. [11, 14, 18, 24, 28, 33, 48, 49, 52, 57, 62, 66, 69, 80, 84, 88, 92, 97, 160, 207]

Hills, D.A., and Sackfield, A. (1986) The stress field induced by a twisting sphere, *Jnl. Appl. Mech.*, **53**, 2, 372-378. [79]

Hills, D.A., and Sackfield, A. (1987) The stress field induced by normal contact between dissimilar elastic spheres, *Jnl. Appl. Mech.*, **54**, 8-14. [77]

Hills, D.A., Waterhouse, R.B., and Noble, B. (1983) An analysis of shot-peening, *Jnl. Strain Anal.*, **18**, 95-100. [123]

Hourlier, F., d'Hondt, H. Truchon, M., and Pineau, A. (1982) Fatigue crack behaviour under polymodal fatigue, Report RE 958, Institut de Recherches de la Sidérurgie Française. [174]

Irwin, G.R. (1957) Analysis of stresses and strains near the end of a crack traversing a plate, *Jnl. Appl. Mech*, **24**, 361-364. [135]

Irwin, G.R. (1960) Plastic zone near a crack and fracture toughness, *Proc. 7th Sagamore Ordnance Materials Conference*, **4**, 63-78, Syracuse University, New York. [174]

Johnson, K.L. (1955) Surface interaction between elastically loaded bodies under tangential forces, *Proc. Roy. Soc., Series A*, **230** , 531-548. [4, 34, 51, 202]

Johnson, K.L. (1983). Letter to editor on paper by Comninou and Barber (1983), *Int. Jnl. Solids Structures*, **19**, 10, 935-936. [120]

Johnson, K.L. (1985) *Contact Mechanics*, Cambridge University Press, Cambridge. [28, 46, 60, 109, 161, 206, 207]

Kalker, J.J. (1990) *Three-Dimensional Elastic Bodies in Rolling Contact*, Kluwer, Dordrecht. [44, 88]

Kanninen, M.F. and Popelar, C.H. (1985) *Advanced Fracture Mechanics*, Oxford University Press, Oxford. [134]

Keer, L.M., Dundurs, J., and Tsai, K.C. (1972) Problems involving receding contact between a layer and a half-space *Jnl. Appl. Mech.*, **39**, 1115-1120. [12]

Khadem, R. and O'Connor, J.J. (1969) Adhesive or frictionless compression of an elastic rectangle between two identical elastic half-spaces, *Int. Jnl. Eng. Sci*, **7**, 153-168. [12, 158]

Kikuchi, N. (1993) *Computational Methods in Contact Mechanics*, Kluwer, Dordrecht. [89]

Komvopoulos, K., and Choi, D.-H. (1992) Elastic finite element analysis of multi-asperity contacts, *Jnl. Tribology*, **114**, 823-831. [105, 112]

Krenk, S. (1975) On the use of the interpolation polynomial for solutions of singular integral equations, *Q. Appl. Maths.*, **32**, 479-484. [139]

Kuno, M., Waterhouse, R.B., Nowell, D., and Hills, D.A. (1989) Initiation and growth of fretting fatigue cracks in the partial slip regime, *Fatigue Fract. Eng. Mats. Structs.*, **12**, 5, 387-398. [49, 161, 190, 202, 203]

Liebensperger, R.L., and Brittain, T.M. (1973) Shear stresses below asperities in Hertzian contact as measured by photoelasticity, *Jnl. Lub. Tech.*, **95**, 277-286. [105]

Lo, C.C. (1969) Elastic contact of rough cylinders, *Int. Jnl. Mech Sci.*, **11**, 105-115. [110]

Love, A.E.H. (1927) *A Treatise on the Mathematical Theory of Elasticity.* Oxford University Press, Oxford. [29]

Love, A.E.H. (1929) The stresses produced in a semi-infinite solid by pressure on part of the boundary, *Phil. Trans. Roy. Soc.*, **A228**, 377-420. [88]

Lubkin, J.L. (1951) The torsion of elastic spheres in contact, *Jnl. Appl. Mech.*, **73**, 183-187. [79, 80]

M'Ewan, E. (1949) Stresses in elastic cylinders in contact along a generatrix, *Phil. Mag.*, **40**, 454-459. [22]

McDowell, J.R. (1953) Fretting corrosion tendencies of several combinations of materials, *ASTM STP 144: Symposium on Fretting Corrosion*, ASTM, Philadelphia. [3]

Miller, K.J. (1991) Metal fatigue - past, current, and future, *Proc. I.Mech.E.*, **C205**, 291-304. [175, 196]

Miller, G.R., Keer, L.M., and Cheng, H.S. (1985) On the mechanics of fatigue crack growth due to contact loading, *Proc. Roy. Soc. London*, **A397**, 197-209. [112]

Mindlin, R.D. (1936) Force at a point in the interior of a semi-infinite solid, *Physics*, **7**, 195-202. [152]

Mindlin, R.D., (1949) Compliance of elastic bodies in contact, *Jnl. Appl. Mech.*, **16**, 259-268. [4, 47, 50, 82, 202]

Mindlin, R.D., and Deresciewicz, H. (1953) Elastic spheres in contact under varying oblique forces, *Jnl. Appl. Mech.*, **75**, 327-344. [58, 60]

Mostofi, A., and Gohar, R. (1980) Pressure distribution between closely conforming contacting surfaces, *Jnl. Mech. Eng. Sci.*, **22** , 5, 251-259. [12]

Munisamy, R.L., Hills, D.A., and Nowell, D. (1992) Contact of similar and dissimilar elastic spheres under tangential loading, pp 447-461 of *Proc. Contact Mechanics International Symposium, Lausanne, Oct 1992*, Ed A. Curnier, Presses Polytechniques et Universitaires Romandes. [52, 58, 72, 76, 202]

Munisamy, R.L., Hills, D.A., and Nowell, D. (1993) Static axisymmetric Hertzian contacts subject to shearing forces, *Jnl. Appl. Mech.*, in press. [76]

Mura, T. (1982) *Micromechanics of Defects in Solids* Martinus Nijhoff, Dordrecht, Netherlands. [149, 152]

Mura, T. and Nakasone, Y. (1990) A theory of fatigue crack initiation in solids, *Jnl. Appl. Mech.*, **57**, 1-6. [123, 131, 210, 212]

Murakami, Y. (1987) (Editor) *Stress Intensity Factors Handbook*, Pergamon Press, Oxford. [149]

Murakami, Y., and Nemat-Nasser, S. (1983) Growth and stability of interacting surface flaws of arbitary shape, *Eng. Fract. Mech.*, **17** , 193- 210. [152]

Muskhelishvili, N.I. (1953) *Some Basic Problems of the Mathematical Theory of Elasticity,* Noordhoff, Gröningen. [18, 135]

Nishioka, K., and Hirakawa, K. (1969) Fundamental investigations of fretting fatigue (part 2), *Bull. JSME*, **12** , 50, 180-187. [199]

Nowell, D. (1988) *An Analysis of Fretting Fatigue*, D.Phil. thesis, University of Oxford. [49, 164, 199, 203]

Nowell, D., and Hills, D.A. (1987a) Mechanics of fretting fatigue tests, *Int. Jnl. Mech. Sci.*, **29**, 5, 355-365. [63]

Nowell, D., and Hills, D.A. (1987b) Open Cracks at or Near Free Edges, *Jnl. Strain Anal.*, **22**, 3, 177-185. [143]

Nowell, D., and Hills, D.A. (1988a) Contact problems incorporating elastic layers, *Int. Jnl. Solids Structures*, **24**, 1, 105-115. [91]

Nowell, D., and Hills, D.A. (1988b) Tractive rolling of tyred cylinders, *Int. Jnl. Mech. Sci.,* **30**, 12, 945-947. [91]

Nowell, D., and Hills, D.A. (1989) Hertzian contact of ground surfaces, *Jnl. Tribology*, **111**, 1, 175-9. [103]

Nowell, D., and Hills, D.A. (1990) Crack initiation criteria in fretting fatigue, *Wear*, **136**, 329-343. [162, 193, 203, 205]

Nowell, D., Hills, D.A., and Dai, D.N. (1993) Energy dissipation and crack initiation in fretting fatigue, *Proc. 20th Leeds-Lyon Symposium on Tribology*, Elsevier, Amsterdam. [212]

Nowell, D., Hills, D.A., and O'Connor, J.J. (1987) An analysis of fretting fatigue, *Proc. Int. Conf. on Tribology*, London, July 1987, I.Mech.E., London. [179]

Nowell, D., Hills, D.A., and Sackfield, A. (1988) Contact of dissimilar elastic cylinders under normal and tangential loading, *Jnl. Mech. Phys. Solids*, **36**, 1, 59-75. [66]

Otsuka, A., Mori, K., and Miyata, T. (1975) The condition of fatigue crack growth in mixed mode condition, *Fracture Mech.*, **7** , 429-439. [192]

Pacelli, M. (1956) Contatto con attrito tra due corpi elastici di forma qualunque: compressione e torsione, *Ann. Schola Norm. Sup. Pisa*, Ser III, **10**, 155-184. [82]

Paris, P.C., Gomez, M.P., and Anderson, W.P. (1961) A rational analytic theory of fatigue, *The trend in engineering*, **13**, 9-14. [131]

Persson, A. (1964) *On the Stress Distribution of Cylindrical Bodies in Elastic Contact,* PhD. dissertation, Chalmers Tekniska Hogskola, Goteborg. [12]

Poritsky, H, (1950) Stresses and deflections of cylindrical bodies in contact with application to contact of gears and of locomotive wheels, *Jnl. Appl. Mech.*, **72** ,191-201. [22, 24, 47, 48]

Radzimovsky, E.I. (1953) Stress distribution and strength condition of two rolling cylinders pressed together, *Univ. of Illinois Eng. Exptl. Station Bull.,* No 408. [22]

Raoof, M. (1993) Free bending fatigue prediction of steel cables with the effect of interwire fretting taken into account, *Proc. Int. Conf. on fretting fatigue, Sheffield, April 1993*, I.Mech.E., London. [78, 203]

Rooke, D.P. and Jones, D.A. (1979) Stress intensity factors in fretting fatigue, *Jnl. Strain Analyis*, **14**, 1-6. [178]

Rooke, D.P., Rayaprolu, D.B., and Aliabadi, M.H. (1992) Crack line and edge Green's functions for stress intensity factors of inclined edge cracks. *Fatigue Fract. Engng. Mater. Struct.*, **15**, 5, 441-461. [136, 178]

Ruiz, C., and Chen, K.C. (1986) Life assessment of dovetail joints between blades and discs in aero-engines, *Proc. Int. Conf. Fatigue, Sheffield*, I.Mech.E., London. [89, 201, 203, 205]

Sackfield, A., and Hills, D.A. (1983) A note on the Hertz contact problem: a correlation of standard formulae, *Jnl. Strain Anal.*, **18** , 3, 195-197. [33]

Sato, K. (1992) Determination and control of contact pressure distribution in fretting fatigue, pp 85-100 of *ASTM STP 1159: Standardisation of fretting fatigue methods and equipment*, Ed. M.H. Attia and R.B. Waterhouse. ASTM, Philadelphia. [89, 160]

Sato, K., Fujii, H., and Kodama, S., (1986) Crack propagation behaviour in fretting fatigue, *Wear*, **107**, 245-262. [181, 182]

Schmueser, D., Comninou, M., and Dundurs, J. (1981) Frictional slip between a layer and a substrate, *Jnl. Engng. Div. ASCE*, **107** , 1103-1118. [147]

Sheppard, S.D., Hills, D.A., and Barber, J.R. (1986) An analysis of fretting cracks - II. Unloading and re-loading phases. *Int. J. Solids Structures*, **22**, 4, 387-396. [147]

Sih, G.C. (1974) Strain energy density factor applied to mixed-mode crack problems, *Int. Jnl. Fracture*, **10**, 305-320. [174]

Smith, J.O., and Liu, C.K. (1953) Stresses due to tangential and normal loads on an elastic solid with application to some contact stress problems, *Jnl. Appl. Mech.*, **21**, 157-166. [22]

Sneddon, I.N., (1951) *Fourier Transforms*, McGraw-Hill, New York. [28, 90, 92]

Soboyejo, W.O. Kishimoto, K., Smith, R.A., and Knott, J.F. (1989) A study of the interaction and coalescence of two coplanar cracks in bending, *Fatigue Fract. Engng Mater. Struct.*, **3**, 167-174. [184]

Spence, D.A. (1973) An eigenvalue problem for elastic contact with finite friction, *Proc. Camb. Phil. Soc.*, 73, 249-268. [67, 68]

Spence, D.A. (1975) The Hertz contact problem with finite friction, *Jnl. Elasticity*, **5**, 3-4, 297-319. [77]

Spink, G.M. (1990) Fretting fatigue of a $2\frac{1}{2}\%$ Ni/Cr/Mo/V low pressure turbine steel - the effect of different contact pad materials and of variable slip amplitude, *Wear*, **136**, 281-297. [199]

Sturt, A., Nowell, D., and Hills, D.A. (1993) Applications of boundary element and dislocation density methods in plane crack problems, *Eng. Anal. with Boundary Elements*, **11**, 129-135. [179]

Suresh, S. (1991) *Fatigue of Materials*, Cambridge University Press, Cambridge. [170, 175, 210]

Timoshenko, S.P., and Goodier, J.N. (1951) *Theory of Elasticity,* McGraw-Hill, New York. [15, 152]

Tomlinson, G. A. (1927) The rusting of steel surfaces in contact, *Proc. Roy. Soc., Series A*, **115**, 472-483. [3]

Toor, P.M. (1975a) On fracture mechanics under complex stress, *Eng. Fract. Mech.*, **7**, 321-329. [127]

Toor, P.M. (1975b) A unified engineering approach to the prediction of multiaxial fatigue of aircraft structures, *Eng. Fract. Mech.*, **7**, 731-741. [127]

Venkataraman, G., Chung, Y-W., Nakasone, Y., and Mura, T. (1990) Free energy formulation of fatigue crack initiation along persistent slip bands: calculation of S-N curves and crack depths, *Acta Metall. Mater.*, **38**, 1, 31-40. [210]

Vingsbo, O., and Soderberg, D. (1988) On fretting maps, *Wear*, **126**, 131-147. [199]

Walker, E.K., (1970) The effect of stress ratio during crack propagation and fatigue for 2024-T3 and 7076-T6 aluminium, in ASTM STP 462: *Effects of Environmental and Complex Load History on Fatigue Life*, ASTM, Philadelphia. [173]

Warlow-Davies, E.J. (1941) Fretting corrosion and fatigue strength: brief results of preliminary experiments, *Proc. I.Mech.E.*, **146** , 33-38. [3]

Waterhouse, R.B. (1972) *Fretting Corrosion*, Pergamon, Oxford. [2, 3]

Waterhouse, R.B. (1981) *Fretting Fatigue*, Applied Science Publishers, London. [2, 3]

Waterhouse, R.B. (1992) Fretting fatigue, *Int. Materials Reviews,* **37**, 2, 77-97. [3]

Waterhouse, R.B., and Lindley, T.C. (1993) *Proceedings of International Conference on Fretting Fatigue*, Sheffield, April 1993, M.E.P., London. [3]

Webster, M.N., and Sayles, R.S. (1986) A numerical model for the elastic frictionless contact of real rough surfaces, *Jnl Tribology*, **108**, 314-320. [111]

West, M.A., and Sayles R.S. (1988) A 3-dimensional method of studying 3-body contact geometry and stress on real rough surfaces, in *Interface Dynamics, Proc 14th Leeds-Lyon Symposium on Tribology*, Ed D. Dowson, C.M. Taylor, M. Godet, D. Berthe, Elsevier, Amsterdam. [111]

Westergaard, H.M. (1939) Bearing pressures and cracks, *Jnl. Appl. Mech.*, **6**, 49-52. [101, 128, 134]

Wilks, M.D.B., Nowell, D., and Hills, D.A. (1993) The evaluation of stress intensity factors for plane cracks in residual stress fields, Jnl. Strain Anal., **28**, 3, 145-152. [124, 148, 150, 177]

Williams, M.L. (1952) Stress singularities resulting from various boundary conditions in angular corners of plates in extension, *Jnl. Appl. Mech.*, **19**, 526-528. [127]

Williams, M.L. (1957) On the stress distribution at the base of a stationary crack, *Jnl. Appl. Mech.*, **24**, 109 - 114. [135]

Wright, G.P. (1970) *Studies in Fretting Fatigue*, D.Phil. thesis, University of Oxford. [4]

Index

Mechanics

From 1990, books on the subject of *mechanics* will be published under two series:
FLUID **MECHANICS AND ITS APPLICATIONS**
Series Editor: R.J. Moreau
SOLID **MECHANICS AND ITS APPLICATIONS**
Series Editor: G.M.L. Gladwell

Prior to 1990, the books listed below were published in the respective series indicated below.

MECHANICS: DYNAMICAL SYSTEMS
Editors: L. Meirovitch and G.Æ. Oravas

1. E.H. Dowell: *Aeroelasticity of Plates and Shells.* 1975 ISBN 90-286-0404-9
2. D.G.B. Edelen: *Lagrangian Mechanics of Nonconservative Nonholonomic Systems.* 1977 ISBN 90-286-0077-9
3. J.L. Junkins: *An Introduction to Optimal Estimation of Dynamical Systems.* 1978 ISBN 90-286-0067-1
4. E.H. Dowell (ed.), H.C. Curtiss Jr., R.H. Scanlan and F. Sisto: *A Modern Course in Aeroelasticity.* *Revised and enlarged edition see under Volume 11*
5. L. Meirovitch: *Computational Methods in Structural Dynamics.* 1980 ISBN 90-286-0580-0
6. B. Skalmierski and A. Tylikowski: *Stochastic Processes in Dynamics.* Revised and enlarged translation. 1982 ISBN 90-247-2686-7
7. P.C. Müller and W.O. Schiehlen: *Linear Vibrations.* A Theoretical Treatment of Multi-degree-of-freedom Vibrating Systems. 1985 ISBN 90-247-2983-1
8. Gh. Buzdugan, E. Mihăilescu and M. Radeş: *Vibration Measurement.* 1986 ISBN 90-247-3111-9
9. G.M.L. Gladwell: *Inverse Problems in Vibration.* 1987 ISBN 90-247-3408-8
10. G.I. Schuëller and M. Shinozuka: *Stochastic Methods in Structural Dynamics.* 1987 ISBN 90-247-3611-0
11. E.H. Dowell (ed.), H.C. Curtiss Jr., R.H. Scanlan and F. Sisto: *A Modern Course in Aeroelasticity.* Second revised and enlarged edition (of Volume 4). 1989 ISBN Hb 0-7923-0062-9; Pb 0-7923-0185-4
12. W. Szemplińska-Stupnicka: *The Behavior of Nonlinear Vibrating Systems.* Volume I: Fundamental Concepts and Methods: Applications to Single-Degree-of-Freedom Systems. 1990 ISBN 0-7923-0368-7
13. W. Szemplińska-Stupnicka: *The Behavior of Nonlinear Vibrating Systems.* Volume II: Advanced Concepts and Applications to Multi-Degree-of-Freedom Systems. 1990 ISBN 0-7923-0369-5
Set ISBN (Vols. 12–13) 0-7923-0370-9

MECHANICS OF STRUCTURAL SYSTEMS
Editors: J.S. Przemieniecki and G.Æ. Oravas

1. L. Frýba: *Vibration of Solids and Structures under Moving Loads.* 1970 ISBN 90-01-32420-2
2. K. Marguerre and K. Wölfel: *Mechanics of Vibration.* 1979 ISBN 90-286-0086-8

Mechanics

3. E.B. Magrab: *Vibrations of Elastic Structural Members.* 1979 ISBN 90-286-0207-0
4. R.T. Haftka and M.P. Kamat: *Elements of Structural Optimization.* 1985
   Revised and enlarged edition see under Solid Mechanics and Its Applications, Volume 1
5. J.R. Vinson and R.L. Sierakowski: *The Behavior of Structures Composed of Composite Materials.* 1986 ISBN Hb 90-247-3125-9; Pb 90-247-3578-5
6. B.E. Gatewood: *Virtual Principles in Aircraft Structures.* Volume 1: Analysis. 1989
   ISBN 90-247-3754-0
7. B.E. Gatewood: *Virtual Principles in Aircraft Structures.* Volume 2: Design, Plates, Finite Elements. 1989 ISBN 90-247-3755-9
   Set (Gatewood 1 + 2) ISBN 90-247-3753-2

MECHANICS OF ELASTIC AND INELASTIC SOLIDS

Editors: S. Nemat-Nasser and G.Æ. Oravas

---

1. G.M.L. Gladwell: *Contact Problems in the Classical Theory of Elasticity.* 1980
   ISBN Hb 90-286-0440-5; Pb 90-286-0760-9
2. G. Wempner: *Mechanics of Solids with Applications to Thin Bodies.* 1981
   ISBN 90-286-0880-X
3. T. Mura: *Micromechanics of Defects in Solids.* 2nd revised edition, 1987
   ISBN 90-247-3343-X
4. R.G. Payton: *Elastic Wave Propagation in Transversely Isotropic Media.* 1983
   ISBN 90-247-2843-6
5. S. Nemat-Nasser, H. Abé and S. Hirakawa (eds.): *Hydraulic Fracturing and Geothermal Energy.* 1983 ISBN 90-247-2855-X
6. S. Nemat-Nasser, R.J. Asaro and G.A. Hegemier (eds.): *Theoretical Foundation for Large-scale Computations of Nonlinear Material Behavior.* 1984 ISBN 90-247-3092-9
7. N. Cristescu: *Rock Rheology.* 1988 ISBN 90-247-3660-9
8. G.I.N. Rozvany: *Structural Design via Optimality Criteria.* The Prager Approach to Structural Optimization. 1989 ISBN 90-247-3613-7

MECHANICS OF SURFACE STRUCTURES

Editors: W.A. Nash and G.Æ. Oravas

---

1. P. Seide: *Small Elastic Deformations of Thin Shells.* 1975 ISBN 90-286-0064-7
2. V. Panc: *Theories of Elastic Plates.* 1975 ISBN 90-286-0104-X
3. J.L. Nowinski: *Theory of Thermoelasticity with Applications.* 1978
   ISBN 90-286-0457-X
4. S. Łukasiewicz: *Local Loads in Plates and Shells.* 1979 ISBN 90-286-0047-7
5. C. Fiřt: *Statics, Formfinding and Dynamics of Air-supported Membrane Structures.* 1983 ISBN 90-247-2672-7
6. Y. Kai-yuan (ed.): *Progress in Applied Mechanics.* The Chien Wei-zang Anniversary Volume. 1987 ISBN 90-247-3249-2
7. R. Negruţiu: *Elastic Analysis of Slab Structures.* 1987 ISBN 90-247-3367-7
8. J.R. Vinson: *The Behavior of Thin Walled Structures.* Beams, Plates, and Shells. 1988
   ISBN Hb 90-247-3663-3; Pb 90-247-3664-1

Mechanics

MECHANICS OF FLUIDS AND TRANSPORT PROCESSES
Editors: R.J. Moreau and G.Æ. Oravas

1. J. Happel and H. Brenner: *Low Reynolds Number Hydrodynamics.* With Special Applications to Particular Media. 1983 ISBN Hb 90-01-37115-9; Pb 90-247-2877-0
2. S. Zahorski: *Mechanics of Viscoelastic Fluids.* 1982 ISBN 90-247-2687-5
3. J.A. Sparenberg: *Elements of Hydrodynamics Propulsion.* 1984 ISBN 90-247-2871-1
4. B.K. Shivamoggi: *Theoretical Fluid Dynamics.* 1984 ISBN 90-247-2999-8
5. R. Timman, A.J. Hermans and G.C. Hsiao: *Water Waves and Ship Hydrodynamics.* An Introduction. 1985 ISBN 90-247-3218-2
6. M. Lesieur: *Turbulence in Fluids.* Stochastic and Numerical Modelling. 1987 ISBN 90-247-3470-3
7. L.A. Lliboutry: *Very Slow Flows of Solids.* Basics of Modeling in Geodynamics and Glaciology. 1987 ISBN 90-247-3482-7
8. B.K. Shivamoggi: *Introduction to Nonlinear Fluid-Plasma Waves.* 1988 ISBN 90-247-3662-5
9. V. Bojarevičs, Ya. Freibergs, E.I. Shilova and E.V. Shcherbinin: *Electrically Induced Vortical Flows.* 1989 ISBN 90-247-3712-5
10. J. Lielpeteris and R. Moreau (eds.): *Liquid Metal Magnetohydrodynamics.* 1989 ISBN 0-7923-0344-X

MECHANICS OF ELASTIC STABILITY
Editors: H. Leipholz and G.Æ. Oravas

1. H. Leipholz: *Theory of Elasticity.* 1974 ISBN 90-286-0193-7
2. L. Librescu: *Elastostatics and Kinetics of Aniosotropic and Heterogeneous Shell-type Structures.* 1975 ISBN 90-286-0035-3
3. C.L. Dym: *Stability Theory and Its Applications to Structural Mechanics.* 1974 ISBN 90-286-0094-9
4. K. Huseyin: *Nonlinear Theory of Elastic Stability.* 1975 ISBN 90-286-0344-1
5. H. Leipholz: *Direct Variational Methods and Eigenvalue Problems in Engineering.* 1977 ISBN 90-286-0106-6
6. K. Huseyin: *Vibrations and Stability of Multiple Parameter Systems.* 1978 ISBN 90-286-0136-8
7. H. Leipholz: *Stability of Elastic Systems.* 1980 ISBN 90-286-0050-7
8. V.V. Bolotin: *Random Vibrations of Elastic Systems.* 1984 ISBN 90-247-2981-5
9. D. Bushnell: *Computerized Buckling Analysis of Shells.* 1985 ISBN 90-247-3099-6
10. L.M. Kachanov: *Introduction to Continuum Damage Mechanics.* 1986 ISBN 90-247-3319-7
11. H.H.E. Leipholz and M. Abdel-Rohman: *Control of Structures.* 1986 ISBN 90-247-3321-9
12. H.E. Lindberg and A.L. Florence: *Dynamic Pulse Buckling.* Theory and Experiment. 1987 ISBN 90-247-3566-1
13. A. Gajewski and M. Zyczkowski: *Optimal Structural Design under Stability Constraints.* 1988 ISBN 90-247-3612-9

Mechanics

MECHANICS: ANALYSIS

Editors: V.J. Mizel and G.Æ. Oravas

1. M.A. Krasnoselskii, P.P. Zabreiko, E.I. Pustylnik and P.E. Sbolevskii: *Integral Operators in Spaces of Summable Functions.* 1976 ISBN 90-286-0294-1
2. V.V. Ivanov: *The Theory of Approximate Methods and Their Application to the Numerical Solution of Singular Integral Equations.* 1976 ISBN 90-286-0036-1
3. A. Kufner, O. John and S. Pučík: *Function Spaces.* 1977 ISBN 90-286-0015-9
4. S.G. Mikhlin: *Approximation on a Rectangular Grid.* With Application to Finite Element Methods and Other Problems. 1979 ISBN 90-286-0008-6
5. D.G.B. Edelen: *Isovector Methods for Equations of Balance.* With Programs for Computer Assistance in Operator Calculations and an Exposition of Practical Topics of the Exterior Calculus. 1980 ISBN 90-286-0420-0
6. R.S. Anderssen, F.R. de Hoog and M.A. Lukas (eds.): *The Application and Numerical Solution of Integral Equations.* 1980 ISBN 90-286-0450-2
7. R.Z. Has'minskiĭ: *Stochastic Stability of Differential Equations.* 1980 ISBN 90-286-0100-7
8. A.I. Vol'pert and S.I. Hudjaev: *Analysis in Classes of Discontinuous Functions and Equations of Mathematical Physics.* 1985 ISBN 90-247-3109-7
9. A. Georgescu: *Hydrodynamic Stability Theory.* 1985 ISBN 90-247-3120-8
10. W. Noll: *Finite-dimensional Spaces.* Algebra, Geometry and Analysis. Volume I. 1987 ISBN Hb 90-247-3581-5; Pb 90-247-3582-3

MECHANICS: COMPUTATIONAL MECHANICS

Editors: M. Stern and G.Æ. Oravas

1. T.A. Cruse: *Boundary Element Analysis in Computational Fracture Mechanics.* 1988 ISBN 90-247-3614-5

MECHANICS: GENESIS AND METHOD

Editor: G.Æ. Oravas

1. P.-M.-M. Duhem: *The Evolution of Mechanics.* 1980 ISBN 90-286-0688-2

MECHANICS OF CONTINUA

Editors: W.O. Williams and G.Æ. Oravas

1. C.-C. Wang and C. Truesdell: *Introduction to Rational Elasticity.* 1973 ISBN 90-01-93710-1
2. P.J. Chen: *Selected Topics in Wave Propagation.* 1976 ISBN 90-286-0515-0
3. P. Villaggio: *Qualitative Methods in Elasticity.* 1977 ISBN 90-286-0007-8

Mechanics

MECHANICS OF FRACTURE
Editors: G.C. Sih

1. G.C. Sih (ed.): *Methods of Analysis and Solutions of Crack Problems.* 1973 ISBN 90-01-79860-8
2. M.K. Kassir and G.C. Sih (eds.): *Three-dimensional Crack Problems.* A New Solution of Crack Solutions in Three-dimensional Elasticity. 1975 ISBN 90-286-0414-6
3. G.C. Sih (ed.): *Plates and Shells with Cracks.* 1977 ISBN 90-286-0146-5
4. G.C. Sih (ed.): *Elastodynamic Crack Problems.* 1977 ISBN 90-286-0156-2
5. G.C. Sih (ed.): *Stress Analysis of Notch Problems.* Stress Solutions to a Variety of Notch Geometries used in Engineering Design. 1978 ISBN 90-286-0166-X
6. G.C. Sih and E.P. Chen (eds.): *Cracks in Composite Materials.* A Compilation of Stress Solutions for Composite System with Cracks. 1981 ISBN 90-247-2559-3
7. G.C. Sih (ed.): *Experimental Evaluation of Stress Concentration and Intensity Factors.* Useful Methods and Solutions to Experimentalists in Fracture Mechanics. 1981 ISBN 90-247-2558-5

MECHANICS OF PLASTIC SOLIDS
Editors: J. Schroeder and G.Æ. Oravas

1. A. Sawczuk (ed.): *Foundations of Plasticity.* 1973 ISBN 90-01-77570-5
2. A. Sawczuk (ed.): *Problems of Plasticity.* 1974 ISBN 90-286-0233-X
3. W. Szczepiński: *Introduction to the Mechanics of Plastic Forming of Metals.* 1979 ISBN 90-286-0126-0
4. D.A. Gokhfeld and O.F. Cherniavsky: *Limit Analysis of Structures at Thermal Cycling.* 1980 ISBN 90-286-0455-3
5. N. Cristescu and I. Suliciu: *Viscoplasticity.* 1982 ISBN 90-247-2777-4

Kluwer Academic Publishers - Dordrecht / Boston / London

Mechanics

SOLID MECHANICS AND ITS APPLICATIONS

Series Editor: G.M.L. Gladwell

Aims and Scope of the Series

The fundamental questions arising in mechanics are: *Why?, How?,* and *How much?* The aim of this series is to provide lucid accounts written by authoritative researchers giving vision and insight in answering these questions on the subject of mechanics as it relates to solids. The scope of the series covers the entire spectrum of solid mechanics. Thus it includes the foundation of mechanics; variational formulations; computational mechanics; statics, kinematics and dynamics of rigid and elastic bodies; vibrations of solids and structures; dynamical systems and chaos; the theories of elasticity, plasticity and viscoelasticity; composite materials; rods, beams, shells and membranes; structural control and stability; soils, rocks and geomechanics; fracture; tribology; experimental mechanics; biomechanics and machine design.

1. R.T. Haftka, Z. Gürdal and M.P. Kamat: *Elements of Structural Optimization.* 2nd rev.ed., 1990 ISBN 0-7923-0608-2
2. J.J. Kalker: *Three-Dimensional Elastic Bodies in Rolling Contact.* 1990 ISBN 0-7923-0712-7
3. P. Karasudhi: *Foundations of Solid Mechanics.* 1991 ISBN 0-7923-0772-0
4. N. Kikuchi: *Computational Methods in Contact Mechanics.* (forthcoming) ISBN 0-7923-0773-9
5. *Not published.*
6. J.F. Doyle: *Static and Dynamic Analysis of Structures.* With an Emphasis on Mechanics and Computer Matrix Methods. 1991 ISBN 0-7923-1124-8; Pb 0-7923-1208-2
7. O.O. Ochoa and J.N. Reddy: *Finite Element Analysis of Composite Laminates.* ISBN 0-7923-1125-6
8. M.H. Aliabadi and D.P. Rooke: *Numerical Fracture Mechanics.* ISBN 0-7923-1175-2
9. J. Angeles and C.S. López-Cajún: *Optimization of Cam Mechanisms.* 1991 ISBN 0-7923-1355-0
10. D.E. Grierson, A. Franchi and P. Riva: *Progress in Structural Engineering.* 1991 ISBN 0-7923-1396-8
11. R.T. Haftka and Z. Gürdal: *Elements of Structural Optimization.* 3rd rev. and exp. ed. 1992 ISBN 0-7923-1504-9; Pb 0-7923-1505-7
12. J.R. Barber: *Elasticity.* 1992 ISBN 0-7923-1609-6; Pb 0-7923-1610-X
13. H.S. Tzou and G.L. Anderson (eds.): *Intelligent Structural Systems.* 1992 ISBN 0-7923-1920-6
14. E.E. Gdoutos: *Fracture Mechanics.* An Introduction. 1993 ISBN 0-7923-1932-X
15. J.P. Ward: *Solid Mechanics.* An Introduction. 1992 ISBN 0-7923-1949-4
16. M. Farshad: *Design and Analysis of Shell Structures.* 1992 ISBN 0-7923-1950-8
17. H.S. Tzou and T. Fukuda (eds.): *Precision Sensors, Actuators and Systems.* 1992 ISBN 0-7923-2015-8
18. J.R. Vinson: *The Behavior of Shells Composed of Isotropic and Composite Materials.* 1993 ISBN 0-7923-2113-8

Kluwer Academic Publishers – Dordrecht / Boston / London

Mechanics

SOLID MECHANICS AND ITS APPLICATIONS

Series Editor: G.M.L. Gladwell

19. H.S. Tzou: *Piezoelectric Shells*. Distributed Sensing and Control of Continua. 1993 ISBN 0-7923-2186-3
20. W. Schiehlen: *Advanced Multibody System Dynamics*. Simulation and Software Tools. 1993 ISBN 0-7923-2192-8
21. C.-W. Lee: *Vibration Analysis of Rotors*. 1993 ISBN 0-7923-2300-9
22. D.R. Smith: *An Introduction to Continuum Mechanics*. 1993 ISBN 0-7923-2454-4
23. G.M.L. Gladwell: *Inverse Problems in Scattering*. An Introduction. 1993 ISBN 0-7923-2478-1
24. G. Prathap: *The Finite Element Method in Structural Mechanics*. 1993 ISBN 0-7923-2492-7
25. J. Herskovits (ed.): *Structural Optimization '93*. 1993 ISBN 0-7923-2510-9
26. M.A. González-Palacios and J. Angeles: *Cam Synthesis*. 1993 ISBN 0-7923-2536-2
27. W.S. Hall: *The Boundary Element Method*. 1993 ISBN 0-7923-2580-X
28. J. Angeles, G. Hommel and P. Kovács (eds.): *Computational Kinematics*. 1993 ISBN 0-7923-2585-0
29. A. Curnier: *Computational Methods in Solid Mechanics*. 1994 ISBN 0-7923-2761-6
30. D.A. Hills and D. Nowell: *Mechanics of Fretting Fatique*. 1994 ISBN 0-7923-2866-3

Kluwer Academic Publishers – Dordrecht / Boston / London

Mechanics

FLUID MECHANICS AND ITS APPLICATIONS

Series Editor: R. Moreau

Aims and Scope of the Series

The purpose of this series is to focus on subjects in which fluid mechanics plays a fundamental role. As well as the more traditional applications of aeronautics, hydraulics, heat and mass transfer etc., books will be published dealing with topics which are currently in a state of rapid development, such as turbulence, suspensions and multiphase fluids, super and hypersonic flows and numerical modelling techniques. It is a widely held view that it is the interdisciplinary subjects that will receive intense scientific attention, bringing them to the forefront of technological advancement. Fluids have the ability to transport matter and its properties as well as transmit force, therefore fluid mechanics is a subject that is particularly open to cross fertilisation with other sciences and disciplines of engineering. The subject of fluid mechanics will be highly relevant in domains such as chemical, metallurgical, biological and ecological engineering. This series is particularly open to such new multidisciplinary domains.

1. M. Lesieur: *Turbulence in Fluids.* 2nd rev. ed., 1990 ISBN 0-7923-0645-7
2. O. Métais and M. Lesieur (eds.): *Turbulence and Coherent Structures.* 1991 ISBN 0-7923-0646-5
3. R. Moreau: *Magnetohydrodynamics.* 1990 ISBN 0-7923-0937-5
4. E. Coustois (ed.): *Turbulence Control by Passive Means.* 1990 ISBN 0-7923-1020-9
5. A.A. Borissov (ed.): *Dynamic Structure of Detonation in Gaseous and Dispersed Media.* 1991 ISBN 0-7923-1340-2
6. K.-S. Choi (ed.): *Recent Developments in Turbulence Management.* 1991 ISBN 0-7923-1477-8
7. E.P. Evans and B. Coulbeck (eds.): *Pipeline Systems.* 1992 ISBN 0-7923-1668-1
8. B. Nau (ed.): *Fluid Sealing.* 1992 ISBN 0-7923-1669-X
9. T.K.S. Murthy (ed.): *Computational Methods in Hypersonic Aerodynamics.* 1992 ISBN 0-7923-1673-8
10. R. King (ed.): *Fluid Mechanics of Mixing.* Modelling, Operations and Experimental Techniques. 1992 ISBN 0-7923-1720-3
11. Z. Han and X. Yin: *Shock Dynamics.* 1993 ISBN 0-7923-1746-7
12. L. Svarovsky and M.T. Thew (eds.): *Hydroclones.* Analysis and Applications. 1992 ISBN 0-7923-1876-5
13. A. Lichtarowicz (ed.): *Jet Cutting Technology.* 1992 ISBN 0-7923-1979-6
14. F.T.M. Nieuwstadt (ed.): *Flow Visualization and Image Analysis.* 1993 ISBN 0-7923-1994-X
15. A.J. Saul (ed.): *Floods and Flood Management.* 1992 ISBN 0-7923-2078-6
16. D.E. Ashpis, T.B. Gatski and R. Hirsh (eds.): *Instabilities and Turbulence in Engineering Flows.* 1993 ISBN 0-7923-2161-8
17. R.S. Azad: *The Atmospheric Boundary Layer for Engineers.* 1993 ISBN 0-7923-2187-1
18. F.T.M. Nieuwstadt (ed.): *Advances in Turbulence IV.* 1993 ISBN 0-7923-2282-7
19. K.K. Prasad (ed.): *Further Developments in Turbulence Management.* 1993 ISBN 0-7923-2291-6
20. Y.A. Tatarchenko: *Shaped Crystal Growth.* 1993 ISBN 0-7923-2419-6

Kluwer Academic Publishers – Dordrecht / Boston / London

Mechanics

FLUID MECHANICS AND ITS APPLICATIONS

Series Editor: R. Moreau

21. J.P. Bonnet and M.N. Glauser (eds.): *Eddy Structure Identification is Free Turbulent Shear Flows.* 1993 ISBN 0-7923-2449-8

Kluwer Academic Publishers – Dordrecht / Boston / London